SpringerBriefs in Electrical and Computer Engineering

Control, Automation and Robotics

Series Editors

Tamer Başar, Coordinated Science Laboratory, University of Illinois at Urbana-Champaign, Urbana, IL, USA

Miroslav Krstic, La Jolla, CA, USA

SpringerBriefs in Control, Automation and Robotics presents concise summaries of theoretical research and practical applications. Featuring compact, authored volumes of 50 to 125 pages, the series covers a range of research, report and instructional content. Typical topics might include:

- a timely report of state-of-the art analytical techniques;
- a bridge between new research results published in journal articles and a contextual literature review;
- a novel development in control theory or state-of-the-art development in robotics;
- an in-depth case study or application example;
- a presentation of core concepts that students must understand in order to make independent contributions; or
- a summation/expansion of material presented at a recent workshop, symposium or keynote address.

SpringerBriefs in Control, Automation and Robotics allows authors to present their ideas and readers to absorb them with minimal time investment, and are published as part of Springer's e-Book collection, with millions of users worldwide. In addition, Briefs are available for individual print and electronic purchase.

Springer Briefs in a nutshell

- 50–125 published pages, including all tables, figures, and references;
- softcover binding;
- publication within 9–12 weeks after acceptance of complete manuscript;
- copyright is retained by author;
- authored titles only – no contributed titles; and
- versions in print, eBook, and MyCopy.

Indexed by Engineering Index.

Publishing Ethics: Researchers should conduct their research from research proposal to publication in line with best practices and codes of conduct of relevant professional bodies and/or national and international regulatory bodies. For more details on individual ethics matters please see: https://www.springer.com/gp/authors-editors/journal-author/journal-author-helpdesk/publishing-ethics/14214

More information about this subseries at http://www.springer.com/series/10198

Hiroyuki Sato

Riemannian Optimization and Its Applications

 Springer

Hiroyuki Sato
Department of Applied Mathematics
and Physics
Kyoto University
Kyoto, Japan

ISSN 2191-8112   ISSN 2191-8120   (electronic)
SpringerBriefs in Electrical and Computer Engineering
ISSN 2192-6786   ISSN 2192-6794   (electronic)
SpringerBriefs in Control, Automation and Robotics
ISBN 978-3-030-62389-0   ISBN 978-3-030-62391-3   (eBook)
https://doi.org/10.1007/978-3-030-62391-3

MATLAB is a registered trademark of The Mathworks, Inc. See https://www.mathworks.com/
trademarks for a list of additional trademarks.

Mathematics Subject Classification (2010): 49M37, 65K05, 90C30, 90C90

This Springer imprint is published by the registered company Springer Nature Switzerland AG
The registered company address is: Gewerbestrasse 11, 6330 Cham, Switzerland

Preface

Mathematical optimization is an important branch of applied mathematics. Different classes of optimization problems are categorized based on their problem structures. While there are many such classes, this book focuses on Riemannian optimization, i.e., optimization on Riemannian manifolds. In addition to its mathematically interesting theory, Riemannian optimization has many applications to real-world problems. This book reviews the basics of Riemannian optimization, introduces algorithms for Riemannian optimization problems (focusing on first-order methods), discusses the theoretical properties of these algorithms, and briefly suggests possible applications of Riemannian optimization to other fields.

In contrast with optimization in Euclidean spaces, Riemannian optimization concerns optimization problems on Riemannian manifolds, which are not linear spaces in general. For example, the Stiefel manifold is a Riemannian manifold obtained by endowing a set of matrices of fixed size and orthonormal columns with a Riemannian structure. Therefore, solving an unconstrained optimization problem on this manifold is equivalent to solving a constrained one in a matrix space, under the constraint that the decision variable matrix has orthonormal columns. This constraint frequently appears in optimization problems related to, e.g., statistical methods such as principal component analysis. Many other examples of Riemannian optimization problems arise in various fields.

The intended readers of this book include engineers familiar with optimization who may be unfamiliar with Riemannian geometry, mathematicians who have knowledge of Riemannian geometry and are interested in optimization, and researchers or students pursuing optimization and geometry or merging of these two fields. Background knowledge of linear algebra and calculus is assumed, but not of mathematical optimization or geometry such as Riemannian manifolds. However, some basic linear algebra and calculus concepts are reviewed to provide the reader with a smooth introduction to Riemannian optimization.

Throughout the book, for many concepts and methods, we start from the Euclidean case and then extend the discussion to the Riemannian case, focusing on the differences between them. To this end, brief reviews of mathematical optimization in Euclidean spaces (called Euclidean optimization) and Riemannian geometry are included. Riemannian optimization is then introduced by merging these concepts. As a developing Riemannian optimization method, the conjugate gradient method on Riemannian manifolds is discussed in detail. Note that the theory and applications are mutually supportive. Therefore, several applications of Riemannian optimization to problems in other fields, e.g., control engineering, are introduced. To provide further perspective, several emerging Riemannian optimization methods (including stochastic optimization methods) are also reviewed.

This book is organized as follows. In Chap. 1, we consider the unconstrained problem of minimizing a function in a Euclidean space, and the problem involving the same objective function on the sphere. The sphere is one of the simplest Riemannian manifolds; thus, the latter case is an example of a Riemannian optimization problem. In subsequent chapters, the concepts and methods introduced in Chap. 1 are explained in detail, generalized to more abstract cases, extended to more sophisticated ones, and applied to problems in other fields. In Chap. 2, we briefly review the basics of some mathematical concepts, such as linear algebra, topological spaces, and Euclidean optimization, as preliminaries to Riemannian optimization. However, because of space limitations, not every concept can be sufficiently explained; thus, we provide references for those readers requiring more detailed information. Readers who are familiar with certain topics in Chap. 2 may skip the corresponding (sub)sections. In Chap. 3, we introduce the concept of Riemannian manifolds and some related topics, along with a basic optimization method on Riemannian manifolds called the steepest descent method. Readers competent with traditional Riemannian geometry may skip the first three sections. In Chap. 4, we discuss the conjugate gradient method (one of the main topics of this book) in Euclidean spaces and its generalization to Riemannian manifolds. In Chap. 5, we discuss applications of Riemannian optimization to fields including numerical linear algebra, considering topics such as the eigenvalue and singular value decompositions; control engineering, using the optimal model reduction as an example; and statistical methods, as in canonical correlation analysis. Chapters 4 and 5 are partly based on the author's works. A brief review of recent developments in the field of Riemannian optimization is provided in Chap. 6, including studies by other authors.

The author expects this book to provide researchers with the basic aspects of Riemannian optimization so that they can apply Riemannian optimization algorithms to their own problems, with the applications introduced herein serving as references.

I would like to express my sincere gratitude to my colleagues. In particular, discussions with Kensuke Aihara, Ellen Hidemi Fukuda, Hiroyuki Kasai, Bamdev Mishra, Takayuki Okuno, Kazuhiro Sato, Yuya Yamakawa, and Xiaojing Zhu were very valuable and constructive toward the development of this manuscript. I am also very grateful to the production team at Springer. The studies by the author reviewed in this book were partly supported by the Kyoto University Hakubi Project and the Kyoto University Foundation.

Kyoto, Japan   Hiroyuki Sato
September 2020

Contents

Chapter 1
Introduction

In situations ranging from everyday life to corporate strategy, one often attempts to minimize or maximize certain indicators under given conditions. A mathematical formulation of such a problem is called a *mathematical optimization problem* or *mathematical programming problem*. In other words, mathematical optimization (or simply *optimization*) involves minimization or maximization of a given real-valued function f with or without constraints on its variables. Here, f and its variables are called the *objective function* and *decision variables*, respectively. The set of all tuples of decision variables satisfying the constraints is called the *feasible set*, wherein each tuple can be considered an available strategy in decision making. Further, because maximizing a function g is equivalent to minimizing $f := -g$, optimization problems can be discussed as minimization problems without loss of generality.

Optimization problems abound in fields such as engineering, physics, economics, statistics, and machine learning, and solving such problems constitutes an important area of research. To formulate a real-world problem as an optimization problem, we must establish an objective function and possibly some constraints. Establishing an optimization problem with a realistic objective function and constraints facilitates decision making if the optimization problem can be solved. However, such a problem may be extremely difficult to solve. An alternative is to establish a relatively simple objective function and constraints by approximating the original real-world problem. The resultant optimization problem may be easy to solve; however, the simple approximation may cause deviation from the original problem. Thus, it is important to model a real-world problem as an optimization problem that captures the essence of the original problem while also being appropriately solvable.

It is also important to discuss how to solve more complex and difficult optimization problems, to widen the scope of efficiently solvable problems. In the current era of big data, the optimization problems that need to be solved are sometimes large-scale. The continuous development of computers and their ever-increasing computational power have made them useful for solving larger and more complex problems. However,

© Springer Nature Switzerland AG 2021
H. Sato, *Riemannian Optimization and Its Applications*,
SpringerBriefs in Control, Automation and Robotics,
https://doi.org/10.1007/978-3-030-62391-3_1

development of optimization theory that can be applied to efficiently solve such difficult problems is essential. Together, computer performance improvement and efficient algorithm development yield greater solving power.

This book focuses on geometric optimization on Riemannian manifolds, called *Riemannian optimization*, which has recently attracted considerable attention. The book (Absil et al. 2008) by Absil, Mahony, and Sepulchre provides a detailed introduction to optimization on manifolds. Whereas in this book, we focus on first-order methods, especially the Riemannian conjugate gradient methods; provide some examples of applications to other fields; and review the recent developments in Riemannian optimization.

Optimization problems are classified according to their properties and, for some of these classes, numerous solving methods have been developed. In *discrete optimization*, which includes integer programming and combinatorial optimization, the variables of the objective function are restricted to discrete values (Nemhuser and Wolsey 1999; Schrijver 2003). In contrast, in *continuous optimization*, we handle continuous variables, which are usually real numbers (Bertsekas 2016; Forst and Hoffmann 2010; Luenberger and Ye 2008; Nocedal and Wright 2006; Rapcsák 1997; Ruszczyński 2006; Snyman 2005). Among continuous optimization problems, problems with a linear objective function and linear equality/inequality constraints are further classified as linear optimization problems, whereas problems involving some nonlinear terms are called nonlinear optimization problems. Riemannian optimization is a relatively new field of research and can be considered a type of nonlinear optimization. A significant feature of Riemannian optimization is that it generally involves optimization on Riemannian manifolds rather than in Euclidean spaces (we call optimization in Euclidean spaces *Euclidean optimization*), i.e., optimization of nonlinear objective functions over nonlinear spaces in general.

The properties of integers or graphs are fundamental to discrete optimization theory. For continuous optimization, discussion of polyhedrons is essential for consideration of linear optimization problems, and calculus in Euclidean spaces is utilized to develop nonlinear Euclidean optimization theory. Similarly, calculus on Riemannian manifolds, which is facilitated by differential geometry (especially by Riemannian geometry), is crucial for optimization on Riemannian manifolds.

In the remainder of this chapter, we focus on unconstrained and constrained minimization problems involving a quadratic function as simple examples of optimization problems. Then, we observe that a problem with the constraint that the sum of the squares of its decision variables should be 1 can be regarded as an unconstrained minimization problem on the unit sphere; this is one of the simplest examples of a Riemannian optimization problem. The steepest descent method on the sphere is also presented as an introduction to Riemannian optimization algorithms.

1.1 Introduction to Riemannian Optimization

In this section, we discuss simple Euclidean and Riemannian optimization problems involving objective functions of the same form and how to solve them. Throughout this section, we define a vector variable $x \in \mathbb{R}^3$ and a constant matrix A as

$$x = \begin{pmatrix} x_1 \\ x_2 \\ x_3 \end{pmatrix} \quad \text{and} \quad A = \begin{pmatrix} 3 & -1 & -1 \\ -1 & 3 & -1 \\ -1 & -1 & 2 \end{pmatrix}. \tag{1.1}$$

1.1.1 Example of Euclidean Optimization

Let us consider a simple *unconstrained optimization problem*, i.e., an optimization problem without constraints, in the three-dimensional Euclidean space $\mathbb{R}^3$.

Problem 1.1 Find $x \in \mathbb{R}^3$ that minimizes the function $f(x) := x^T A x$ with A in (1.1).[1]

Remark 1.1 From the linear algebra perspective, A is a symmetric positive definite matrix. Hence, $f(x) > 0$ for any $x \neq 0$ and $f(0) = 0$, implying that $x = 0$ is the unique *global optimal solution* to Problem 1.1, i.e., the global minimum point of f.[2] Our aim is to derive an algorithm to solve this type of problem numerically, under the assumption that the solution is unknown. The derived algorithm can also be applied to a larger-scale problem in which the objective function has additional linear terms, which is nontrivial and worth solving. Specifically, if $f(x) = x^T A x - 2b^T x$, having the additional linear term $-2b^T x$ where b is a constant vector, the optimal solution can be written as $A^{-1}b$. This relates the optimization problem to a system of linear equations, i.e., $Ax = b$ (see Sect. 4.1). We consider f with no linear term in Problem 1.1, to unify the form of the objective functions in Problems 1.1 and 1.2 (see the next subsection).

A simple idea for seeking the solution is that, at (x, y) on the graph of $y = f(x)$, we move x in the "steepest downhill" direction. First, we consider the direction $d = (d_1, d_2, d_3)^T \in \mathbb{R}^3$ in which the value of f shows the maximum decrease for a given vector $x \in \mathbb{R}^3$. Accordingly, we temporarily fix the standard Euclidean norm of d to 1, i.e., $\|d\| := \sqrt{d_1^2 + d_2^2 + d_3^2} = 1$. The first-order derivative $g_d'(0)$ of the function $g_d(t) := f(x + td)$ of one variable t at $t = 0$ is then computed as

---

[1] Here, $\cdot^T$ denotes the transposition. See Sect. 1.2 for the notation used in this book.

[2] The fact that $x = 0$ is the unique global optimal solution is also easily verified by rewriting f as $f(x) = 3x_1^2 + 3x_2^2 + 2x_3^2 - 2x_1x_2 - 2x_2x_3 - 2x_3x_1 = (x_1 - x_2)^2 + (x_2 - x_3)^2 + (x_3 - x_1)^2 + x_1^2 + x_2^2$.

$$g'_d(0) = 2(3x_1 - x_2 - x_3)d_1 + 2(-x_1 + 3x_2 - x_3)d_2 + 2(-x_1 - x_2 + 2x_3)d_3$$

$$= \frac{\partial f}{\partial x_1}(\boldsymbol{x})d_1 + \frac{\partial f}{\partial x_2}(\boldsymbol{x})d_2 + \frac{\partial f}{\partial x_3}(\boldsymbol{x})d_3 = \nabla f(\boldsymbol{x})^T \boldsymbol{d} = (2A\boldsymbol{x})^T \boldsymbol{d}.$$

Here, $\frac{\partial f}{\partial x_i}$, $i = 1, 2, 3$ are the partial derivatives of f with respect to x_i, and the vector $\nabla f(\boldsymbol{x}) := \left(\frac{\partial f}{\partial x_1}(\boldsymbol{x}), \frac{\partial f}{\partial x_2}(\boldsymbol{x}), \frac{\partial f}{\partial x_3}(\boldsymbol{x}) \right)^T$ is the *(Euclidean) gradient* of f at $\boldsymbol{x}$. We have $\nabla f(\boldsymbol{x}) = 2A\boldsymbol{x}$ in this case (see also Sect. 2.6 for details). If $\nabla f(\boldsymbol{x}) = \boldsymbol{0}$, then $g'_d(0) = 0$ for any $\boldsymbol{d}$. We assume $\nabla f(\boldsymbol{x}) \neq \boldsymbol{0}$ in the following.[3]

We endow $\mathbb{R}^3$ with the standard inner product $\langle \cdot, \cdot \rangle$ defined by

$$\langle \boldsymbol{a}, \boldsymbol{b} \rangle := \boldsymbol{a}^T \boldsymbol{b} = a_1 b_1 + a_2 b_2 + a_3 b_3 \tag{1.2}$$

for arbitrary $\boldsymbol{a} = (a_1, a_2, a_3)^T$, $\boldsymbol{b} = (b_1, b_2, b_3)^T \in \mathbb{R}^3$. Then, we obtain the expression $g'_d(0) = \langle \nabla f(\boldsymbol{x}), \boldsymbol{d} \rangle$, by which $g'_d(0)$ attains its minimum value when the vector $\boldsymbol{d}$ with unit norm is in the direction opposite to $\nabla f(\boldsymbol{x})$, i.e., $\boldsymbol{d} = -\nabla f(\boldsymbol{x}) / \|\nabla f(\boldsymbol{x})\|$, noting that we assume $\nabla f(\boldsymbol{x}) \neq \boldsymbol{0}$. Hence, we observe that the value of f exhibits almost the maximum decrease when we move $\boldsymbol{x}$ in this direction, assuming that the length of the move is fixed and sufficiently small. In this sense, the direction of $-\nabla f(\boldsymbol{x})$ is called the *steepest descent direction* of f at $\boldsymbol{x}$.

We now seek a point at which the value of f is smaller than $f(\boldsymbol{x})$, by moving $\boldsymbol{x}$ in the direction of $-\nabla f(\boldsymbol{x}) / \|\nabla f(\boldsymbol{x})\|$. The direction chosen to seek the next point $\boldsymbol{x}_+$ from the current point $\boldsymbol{x}$ is called a *search direction* at $\boldsymbol{x}$. Since $-\nabla f(\boldsymbol{x}) / \|\nabla f(\boldsymbol{x})\|$ and $-\nabla f(\boldsymbol{x})$ are in the same direction if the gradient vector $\nabla f(\boldsymbol{x})$ is not $\boldsymbol{0}$, let the search direction be $\boldsymbol{d} := -\nabla f(\boldsymbol{x})$ and the next point be $\boldsymbol{x}_+ := \boldsymbol{x} + t\boldsymbol{d}$. Here, the real value $t > 0$, called a *step length*, determines how far the next point is from the current point. Note that it is not necessarily true that the farther we move from $\boldsymbol{x}$ in the direction of $-\nabla f(\boldsymbol{x})$, the greater the decrease in f.

Therefore, it is necessary to find an appropriate value of t. This procedure is called *line search*. If the objective function is sufficiently simple, we can analytically find $t > 0$ that minimizes $g_d(t) = f(\boldsymbol{x} + t\boldsymbol{d})$. This t is called the step length by *exact line search*. Since $g_d(t)$ is a quadratic function of t for Problem 1.1, exact line search can be performed to obtain

$$t = -\frac{\langle \nabla f(\boldsymbol{x}), \boldsymbol{d} \rangle}{2f(\boldsymbol{d})} \tag{1.3}$$

by noting the relationship

---

[3] In fact, $\nabla f(\boldsymbol{x}) (= 2A\boldsymbol{x}) = \boldsymbol{0}$ is equivalent to $\boldsymbol{x} = \boldsymbol{0}$ in Problem 1.1 because A is invertible. Thus, if $\boldsymbol{x}$ is not equal to the global optimal solution $\boldsymbol{0}$, we can conclude that $\nabla f(\boldsymbol{x}) \neq \boldsymbol{0}$. This means that $\boldsymbol{0}$ is the only local (and, in fact, global) optimal solution to Problem 1.1 (see Theorem 2.2).

Table 1.1 Results of steepest descent method in $\mathbb{R}^3$ for Problem 1.1

k	$\left\|x^{(k)}\right\|$	$f\left(x^{(k)}\right)$	$\left\|\nabla f\left(x^{(k)}\right)\right\|$
0	1.000	6.116×10^{-1}	1.336
1	5.569×10^{-1}	2.572×10^{-1}	1.278
2	4.205×10^{-1}	1.081×10^{-1}	5.618×10^{-1}
5	9.847×10^{-2}	8.041×10^{-3}	2.260×10^{-1}
10	1.315×10^{-2}	1.057×10^{-4}	1.757×10^{-2}
20	1.729×10^{-4}	1.828×10^{-8}	2.310×10^{-4}
40	2.989×10^{-8}	5.463×10^{-16}	3.993×10^{-8}

$$g_d(t) = f(x + td) = (x + td)^T A(x + td) = f(d) t^2 + \langle \nabla f(x), d \rangle t + f(x) \tag{1.4}$$

and assuming that $f(d)$, which is the coefficient of t^2 in (1.4), is positive.[4] However, it is often difficult to perform exact line search for a general objective function. Moreover, exact line search is not necessarily required since it may be sufficient to find $t > 0$ that decreases $g_d(t)$ "to a certain extent" (see Sect. 2.9.1 for further details on line search).

Let k be an integer not smaller than 0. The *steepest descent method* iteratively computes the next point $x^{(k+1)}$ using the search direction $d^{(k)} := -\nabla f\left(x^{(k)}\right)$ and an appropriately determined step length $t^{(k)}$ from the current point $x^{(k)}$ as

$$x^{(k+1)} := x^{(k)} + t^{(k)} d^{(k)} = x^{(k)} - t^{(k)} \nabla f\left(x^{(k)}\right).$$

We execute this algorithm with the initial point $x^{(0)} = (6/11, 6/11, 7/11)^T$ by computing the step lengths (1.3), i.e., through exact line search. Table 1.1 summarizes the numerical results. From the table, $x^{(k)}$ approaches the optimal solution $\mathbf{0}$ as the iterations proceed. We can also observe that $f\left(x^{(k)}\right)$ approaches its minimum value $f(\mathbf{0}) = 0$ and that $\left\|\nabla f\left(x^{(k)}\right)\right\|$ approaches $\left\|\nabla f(\mathbf{0})\right\| = \|\mathbf{0}\| = 0$.

1.1.2 Example of Riemannian Optimization

Next, we consider a problem similar to Problem 1.1 from the previous subsection.

Problem 1.2 Find $x \in \mathbb{R}^3$ that minimizes the function $f(x) := x^T A x$ subject to $x^T x = 1$, where A is defined in (1.1).

The only difference between Problems 1.1 and 1.2—and it is a significant difference— is that Problem 1.2 has the constraint $x^T x = 1$, i.e., $x_1^2 + x_2^2 + x_3^2 = 1$. This problem can also be analytically solved. In fact, because $x = (x_1, x_2, x_3)^T$ can be considered a point on the unit sphere, we can write $x_1 = \sin \theta \cos \phi$, $x_2 = \sin \theta \sin \phi$, and

---

[4] As discussed in Remark 1.1, $f(d) > 0$ holds true for any $d \neq \mathbf{0}$.

$x_3 = \cos\theta$ with $0 \leq \theta \leq \pi$ and $0 \leq \phi < 2\pi$. Then, $f(x)$ restricted to the sphere can be regarded as the function $F(\theta, \phi) := f(\sin\theta\cos\phi, \sin\theta\sin\phi, \cos\theta)$ of two variables θ and ϕ, which is evaluated as

$$F(\theta, \phi) = 2 + \sin^2\theta(1 - \sin 2\phi) - \sqrt{2}\sin 2\theta \sin\left(\phi + \frac{\pi}{4}\right) \geq 2 - \sqrt{2}.$$

The equals sign in the inequality holds true if and only if $\sin^2\theta(1 - \sin 2\phi) = 0$ and $\sin 2\theta \sin(\phi + \pi/4) = 1$, which is equivalent to $(\theta, \phi) = (\pi/4, \pi/4), (3\pi/4, 5\pi/4)$. Therefore, the optimal solutions to Problem 1.2 are $x = \pm(1/2, 1/2, 1/\sqrt{2})^T$ and the minimum value of f is $2 - \sqrt{2}$.

Optimization problems such as Problem 1.2, which include constraints on the decision variables, are called *constrained optimization problems*. In the above discussion, we rewrote this problem, which has three real variables $x_1, x_2, x_3 \in \mathbb{R}$ and a constraint, as an unconstrained problem with two variables $\theta, \phi \in \mathbb{R}$, by appropriately handling the constraint. Note that Problem 1.2 is artificially created so that it can be solved through manual calculation. Practically, it is often more difficult to exploit such a structure (i.e., to reduce the number of variables) in the case of a larger-scale problem, which must then be solved computationally. Such a constrained optimization problem is sometimes solved by additionally introducing Lagrange multipliers associated with the constraints as new variables. However, one may wonder whether the problem could be solved by reducing the dimension of the original problem instead of adding variables, because the degrees of freedom of the variables should be reduced by the constraints. Geometric optimization on Riemannian manifolds, or *Riemannian optimization*, explores this approach.

Similar to the previous subsection, we consider an optimization algorithm for Problem 1.2 as an example of Riemannian optimization. If we perform line search in the direction of $-\nabla f(x)$, as for Problem 1.1, the constraint is no longer satisfied because $-\nabla f$ has no information on the constraint. A key idea in the algorithm below is that we seek the solution to Problem 1.2 from the set of all points satisfying $x^T x = 1$, i.e., the unit *sphere* $S^2 := \{x \in \mathbb{R}^3 \mid \|x\| = 1\}$. In other words, we regard Problem 1.2 as the following *unconstrained optimization problem* on S^2.

Problem 1.3 Find $x \in S^2$ that minimizes the function $f(x) := x^T A x$ with A in (1.1).

Strictly speaking, this f is the restriction of f in Problem 1.2 to S^2. For simplicity, we use the same notation f for both functions although their domains differ.

The direction of $\nabla f(x)$ has no relation to S^2. Thus, we project $\nabla f(x)$ to the tangent plane $T_x S^2$, which is the set of all *tangent vectors* of S^2 at x and is a subspace of $\mathbb{R}^3$ containing $\mathbf{0}$. In other words, $T_x S^2$ is obtained by parallel translation of the plane tangent to S^2 at x as an affine space containing x, such that x is moved to the origin of $\mathbb{R}^3$. Specifically, $T_x S^2$ is expressed as $T_x S^2 = \{\xi \in \mathbb{R}^3 \mid x^T \xi = 0\}$ (see Sects. 3.1.2 and 3.1.4 for details and stricter treatment of tangent spaces). We now endow $T_x S^2$ with the inner product (1.2); i.e., for arbitrary tangent vectors $\xi, \eta \in T_x S^2 \subset \mathbb{R}^3$, we define $\langle \xi, \eta \rangle := \xi^T \eta$. The orthogonal projection P_x onto $T_x S^2$ is given by

Fig. 1.1 Sphere S^2 and tangent space $T_x S^2$ at point $x \in S^2$. A tangent vector at x is regarded as a vector in $\mathbb{R}^3$ with foot x. The Euclidean gradient $\nabla f(x)$ with foot x is not in $T_x S^2$ in general. To make it tangent to S^2, the orthogonal projection P_x is used to obtain grad $f(x) \in T_x S^2$

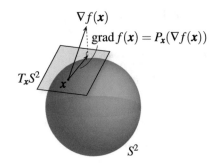

$$P_x(a) = a - \langle x, a \rangle x$$

for any $a \in \mathbb{R}^3$. Let grad $f(x)$ denote the projected vector of $\nabla f(x)$:

$$\text{grad } f(x) = P_x(\nabla f(x)). \tag{1.5}$$

As discussed in Chap. 3, this is the *Riemannian gradient* of f on the Riemannian manifold S^2, endowed with the induced Riemannian metric from the standard inner product $\langle \cdot, \cdot \rangle$ in $\mathbb{R}^3$. Figure 1.1 summarizes these concepts.

Similar to the case of Problem 1.1, let the *search direction* d be $-$ grad $f(x)$ and suppose that we have an appropriate step length $t > 0$. We again assume $d \neq 0$. Nevertheless, the next point cannot be computed as $x + td$ since it is not on S^2. One way to compute the next point on S^2 is to consider the great circle $\gamma(t)$ on S^2 such that $\gamma(0) = x$ and $\dot{\gamma}(0) := \frac{d}{dt}\gamma(t)\big|_{t=0} = d$, which is analogous to a straight line in a Euclidean space. Let $e_1 := (1, 0, 0)^T$ and $e_2 := (0, 1, 0)^T$. The great circle

$$\hat{\gamma}(t) := (\cos(\|d\|t), \sin(\|d\|t), 0)^T = \cos(\|d\|t)e_1 + \sin(\|d\|t)e_2,$$

satisfying $\hat{\gamma}(0) = e_1$ and $\dot{\hat{\gamma}}(0) = \|d\|e_2$, with the 3×3 rotation matrix Q such that $Qx = e_1$ and $Qd = \|d\|e_2$ can be used to give the following formula for $\gamma(t)$:

$$\gamma(t) = Q^{-1}\hat{\gamma}(t) = \cos(\|d\|t)x + \frac{\sin(\|d\|t)}{\|d\|}d. \tag{1.6}$$

The next point can then be computed as (1.6) instead of $x + td$. In the case of $\|d\|(t_2 - t_1) \in [0, \pi]$, $\gamma([t_1, t_2])$ is the shortest curve among the curves connecting the two points $\gamma(t_1)$ and $\gamma(t_2)$, which is called the *geodesic* in a general case. This is a natural extension of line search in $\mathbb{R}^3$ to the sphere S^2. However, if we simply want to retract $x + td \in \mathbb{R}^3$, which is outside S^2, to S^2, an easier approach is to compute the next point as $(x + td)/\|x + td\|$, which is the normalization of $x + td$.

Summarizing the above, the *(Riemannian) steepest descent method* on the sphere S^2 generates a sequence $\{x^{(k)}\}$ on S^2 as follows. Given a point $x^{(k)} \in S^2$ at the

Fig. 1.2 Retraction R
retracting point
$x - \operatorname{grad} f(x)$ to S^2. Here,
R_x is the restriction of R to
$T_x S^2$. The second choice
in (1.7) is adopted.
Collecting the points
$R_x(-t \operatorname{grad} f(x))$ on S^2 for
$t \in [0, \infty)$, we obtain the
curve
$\{R_x(-t \operatorname{grad} f(x)) \mid t \in [0, \infty)\}$

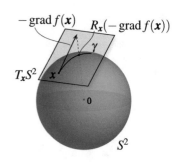

kth iteration and using the search direction $d^{(k)} := -\operatorname{grad} f(x^{(k)}) \in T_x S^2$ and an appropriate step length $t^{(k)} > 0$, the next point $x^{(k+1)}$ is computed as

$$x^{(k+1)} := R_{x^{(k)}}(t^{(k)} d^{(k)}) = R_{x^{(k)}}(-t^{(k)} \operatorname{grad} f(x^{(k)}))$$
$$= R_{x^{(k)}}(-t^{(k)} P_{x^{(k)}}(\nabla f(x^{(k)}))),$$

where R_x is a map from $T_x S^2$ to S^2 defined as

$$R_x(d) := (\cos \|d\|)x + \frac{\sin \|d\|}{\|d\|} d \quad \text{or} \quad R_x(d) := \frac{x+d}{\|x+d\|}. \qquad (1.7)$$

We can search for an appropriate value of $t^{(k)}$ using the curve $\gamma_k(t) := R_{x^{(k)}}(t d^{(k)})$. The concepts discussed above are illustrated in Fig. 1.2.

Remark 1.2 There are other appropriate maps R_x besides those in (1.7), which can be generally explained using the concept of a *retraction*, introduced in Sect. 3.4. We compare γ, defined by (1.6), with $\tilde{\gamma}(t) := (x + td)/\|x + td\|$ to observe that both curves satisfy $\gamma(0) = \tilde{\gamma}(0) = x$ and $\dot{\gamma}(0) = \dot{\tilde{\gamma}}(0) = d$. Thus, $\tilde{\gamma}$ is considered a first-order approximation of the geodesic γ. Although the images of $[0, \pi/(2\|d\|))$ by γ and $[0, \infty)$ by $\tilde{\gamma}$ coincide in this case, images by different retractions on a general manifold may neither coincide nor have an inclusive relation.

Figure 1.3 illustrates the graph of $f(x)$ on the sphere S^2. Here, for each $x \in S^2$, we plot y in the radial direction such that the distance between x and y is equal to $f(x)$, i.e., $y = x + f(x)x = (1 + f(x))x \in \mathbb{R}^3$. Table 1.2 summarizes the numerical results of the Riemannian steepest descent method for Problem 1.3 with the initial point $x^{(0)} = (6/11, 6/11, 7/11)^T \in S^2$ and fixed $t^{(k)} \equiv 0.1$, where $x^* := (1/2, 1/2, 1/\sqrt{2})^T$ is an optimal solution and $f(x^*) = 2 - \sqrt{2}$ is the minimum value of f on S^2. Although a step length of 0.1 does not sufficiently reflect the information on the objective function, the generated sequence converges to an optimal solution for this example. Searching on curves defined by a retraction and selection of more meaningful step lengths are both discussed in Sect. 3.5.

The above problem is generalized by considering an optimization problem on the higher-dimensional $(n-1)$-(hyper)sphere $S^{n-1} := \{x \in \mathbb{R}^n \mid \|x\| = 1\}$, where $\|\cdot\|$ is the Euclidean norm and $\|x\| = 1$ is equivalent to $x^T x = 1$. This book provides

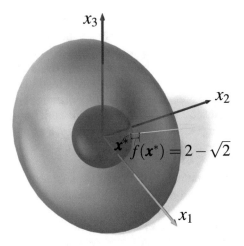

Fig. 1.3 Unit sphere S^2 in $\mathbb{R}^3$ and graph of $f(x)$ on S^2, which is expressed as $\{y = (1 + f(x))x \mid x \in S^2\}$. Here, the graph is illustrated as a surface located outside S^2 since $f(x) > 0$ for all $x \in S^2$. One of the optimal solutions, $x^* = (1/2, 1/2, 1/\sqrt{2})$, is indicated in the figure; at this point, the corresponding $y^* = (1 + f(x^*))x^*$ is one of the most recessed points on the surface. For information, the yellow line is the half-line emanating from the origin in the direction of x^*

Table 1.2 Results of steepest descent method on sphere S^2 with fixed step length of 0.1 for Problem 1.3

k	$\left\| x^{(k)} - x^* \right\|$	$f(x^{(k)}) - f(x^*)$	$\left\| \operatorname{grad} f(x^{(k)}) \right\|$
0	9.559×10^{-2}	2.578×10^{-2}	5.376×10^{-1}
1	4.191×10^{-2}	4.965×10^{-3}	2.368×10^{-1}
2	1.823×10^{-2}	9.404×10^{-4}	1.031×10^{-1}
5	1.494×10^{-3}	6.317×10^{-6}	8.454×10^{-3}
10	2.309×10^{-5}	1.509×10^{-9}	1.306×10^{-4}
15	3.569×10^{-7}	3.604×10^{-13}	2.019×10^{-6}
20	5.515×10^{-9}	2.220×10^{-16}	3.120×10^{-8}

algorithms for minimizing objective functions defined on more general Riemannian manifolds. Because a Euclidean space is the simplest example of a Riemannian manifold, Euclidean optimization can be considered a special case of Riemannian optimization. However, we should remember that our extension of the discussion to Riemannian manifolds is based on the findings of numerous studies on Euclidean optimization. Thus, in the next chapter, we briefly review Euclidean optimization, as well as mathematical preliminaries, for use in Chap. 3, where we discuss the fundamental concepts of Riemannian optimization.

Although we have presented an overview of geometric optimization on the sphere, we have not discussed some issues pertaining to general Riemannian optimization, as detailed below.

- We discussed the steepest descent method, which is one of the most basic optimization algorithms. However, we must ask whether the steepest descent direction is the best search direction. In other words, are there better search directions? The steepest descent method constitutes a basic concept that aids development of other optimization algorithms, and many optimization algorithms have been already proposed in the fields of both Euclidean and Riemannian optimization. In the following chapters, we discuss a more general framework for Riemannian optimization methods, including the steepest descent method, and introduce state-of-the-art algorithms such as the conjugate gradient (Chap. 4) and stochastic optimization methods (Sect. 6.1).
- We discussed an optimization problem on the sphere as an example. However, we must investigate optimization for a problem defined on a more general Riemannian manifold. In Chap. 3, we introduce the concept of Riemannian manifolds, including the (hyper)sphere as a special case, and discuss optimization on general Riemannian manifolds. We also discuss some important examples of Riemannian manifolds, such as the Stiefel and Grassmann manifolds.
- The optimization problem on the sphere discussed in this section is related to an eigenvalue problem. There are many other applications of Riemannian optimization to other fields such as control theory and statistics. In Chap. 5, we discuss such applications together with specific Riemannian optimization problems, i.e., the specific form of the objective function and the Riemannian manifold for each application.

We conclude this section with a remark on one additional advantage of Riemannian optimization. As shown in the example in Sect. 1.1.2, an optimization problem on a Riemannian manifold defined by the original constraints of the Euclidean problem can be regarded as a problem where we seek an optimal solution on the manifold whose dimension is smaller (at least not larger) than that of the Euclidean space considered in the original problem. In particular, each iteration of Riemannian Newton's method requires solving a linear system on a tangent space called Newton's equation, which is a computationally expensive task. Because the dimension of each tangent space is equal to the dimension d of the manifold, the dimension of the linear system to be solved is also d. Therefore, a smaller value of d yields a less expensive iteration of Newton's method, which is an advantage of considering a manifold of smaller dimension (see Sect. 6.3.1 for details).

1.2   Notation

Hereafter, numerical vectors are denoted by regular lowercase letters (not boldface). Matrices are denoted by uppercase letters. Throughout the book, numerical vectors are basically handled as column vectors, i.e., an n-dimensional numerical vector is a vertical arrangement of n numbers. The sets of all real numbers, complex numbers, and natural numbers (nonnegative integers), are denoted by $\mathbb{R}, \mathbb{C}$, and $\mathbb{N}$, respectively.

The vector spaces of all n-dimensional real numerical vectors and $m \times n$ real matrices are denoted by $\mathbb{R}^n$ and $\mathbb{R}^{m \times n}$, respectively, and $\mathbb{R}^n$ is identified with $\mathbb{R}^{n \times 1}$. A numerical vector $x \in \mathbb{R}^n$ whose ith element is x_i is written as $x = (x_i)$. Similarly, if the (i, j) element of a matrix $A \in \mathbb{R}^{m \times n}$ is a_{ij}, then A is written as $A = (a_{ij})$. We also denote the (i, j) element of A by A_{ij}. For complex cases, we use the same notation but replace $\mathbb{R}$ with $\mathbb{C}$.

A member of a sequence $\{x_k\}$ in a set M is denoted by x_k (in the concrete examples considered in this book, this member can usually be expressed as a numerical vector or matrix), which is simpler than the $x^{(k)}$ notation used in the previous section. When we need to distinguish the ith element $x_i \in \mathbb{R}$ of $x \in \mathbb{R}^n$ from the kth member $x_k \in M$ of a sequence $\{x_k\} \subset M$, we write the ith element of x as $(x)_i$.

For sets X and Y, let $X - Y$ and $X \times Y$ denote the relative complement of Y in X and the product set of X and Y, respectively, i.e., $X - Y := \{x \in X \mid x \notin Y\}$ and $X \times Y := \{(x, y) \mid x \in X, y \in Y\}$. The identity map on X, which maps any $x \in X$ to x, is denoted by id_X. If the domain X is clear from the context, we also write id.

The Kronecker delta is denoted by δ_{ij}, i.e., $\delta_{ij} = 1$ when $i = j$, and $\delta_{ij} = 0$ otherwise, where i and j are integers. The identity matrix of order n is denoted by I_n, i.e., $I_n = (\delta_{ij})$. The transpose of a matrix $A = (a_{ij}) \in \mathbb{R}^{m \times n}$ is denoted by A^T, and if $m = n$, the trace and inverse of A are denoted by $\mathrm{tr}(A)$ and A^{-1}, respectively, i.e., $A^T = (a_{ji})$, $\mathrm{tr}(A) = \sum_{i=1}^{n} a_{ii}$, and $AA^{-1} = A^{-1}A = I_n$. If A^{-1} exists for a square matrix A, it holds that $(A^{-1})^T = (A^T)^{-1}$, where the quantity on both sides is denoted by A^{-T} for simplicity. We use the notation $\mathrm{diag}(d_1, d_2, \ldots, d_n)$ with $d_1, d_2, \ldots, d_n \in \mathbb{R}$ to denote the $n \times n$ diagonal matrix whose ith diagonal element is d_i.

The set of all $n \times n$ symmetric (resp. skew-symmetric) matrices is denoted by $\mathrm{Sym}(n)$ (resp. $\mathrm{Skew}(n)$), and the orthogonal group of order n, which is the set of all $n \times n$ orthogonal matrices, is denoted by $\mathcal{O}_n$, i.e., $\mathrm{Sym}(n) := \{A \in \mathbb{R}^{n \times n} \mid A = A^T\}$, $\mathrm{Skew}(n) := \{A \in \mathbb{R}^{n \times n} \mid A = -A^T\}$, and $\mathcal{O}_n := \{A \in \mathbb{R}^{n \times n} \mid A^T A = A A^T = I_n\}$. For a square matrix A, the symmetric (resp. skew-symmetric) part of A is denoted by $\mathrm{sym}(A)$ (resp. $\mathrm{skew}(A)$), i.e., $\mathrm{sym}(A) = (A + A^T)/2$ and $\mathrm{skew}(A) = (A - A^T)/2$.

Chapter 2
Preliminaries and Overview of Euclidean Optimization

In this chapter, we briefly review some mathematical concepts as preliminaries for our discussion of Riemannian optimization; these concepts range from linear algebra to general topology and include Euclidean optimization. Although readers are expected to have basic knowledge of linear algebra and calculus, we review some aspects of these topics in consideration of their relationship to manifolds. One of our goals is to establish a foundation for discussion of Riemannian optimization; therefore, throughout this chapter, we note important information where appropriate.

2.1 Basic Matrix Properties

For matrices A and B such that the product AB is defined, we have $(AB)^T = B^T A^T$. Further, if both AB and BA are defined, $\text{tr}(AB) = \text{tr}(BA)$ holds true. These properties are frequently used throughout the book.

For a matrix $A \in \mathbb{R}^{n \times n}$, if $\lambda \in \mathbb{C}$ and $x \in \mathbb{C}^n - \{0\}$ satisfy $Ax = \lambda x$, then λ and x are called an *eigenvalue* of A and an *eigenvector* of A associated with λ, respectively. Each eigenvalue of a symmetric matrix $A \in \text{Sym}(n)$ is a real number, and $A \in \text{Sym}(n)$ is said to be *positive semidefinite* if $x^T A x \geq 0$ for all $x \in \mathbb{R}^n$. A symmetric positive semidefinite matrix A is said to be *positive definite* if $x^T A x > 0$ for all nonzero $x \in \mathbb{R}^n$. A symmetric matrix A is positive semidefinite (resp. positive definite) if and only if all eigenvalues of A are nonnegative (resp. positive).

There are various types of matrix decomposition, which usually depend on the matrix characteristics. For more detailed discussions, refer to Golub and Van Loan (2013), Meyer (2000).

Proposition 2.1 *For any symmetric positive definite matrix $A \in \text{Sym}(n)$, there exists a unique lower triangular matrix $L \in \mathbb{R}^{n \times n}$ for which the diagonal elements are all positive such that A can be decomposed into $A = LL^T$.*

© Springer Nature Switzerland AG 2021
H. Sato, *Riemannian Optimization and Its Applications*,
SpringerBriefs in Control, Automation and Robotics,
https://doi.org/10.1007/978-3-030-62391-3_2

Proposition 2.2 *For any matrix $A \in \mathbb{R}^{m \times n}$ with $m \geq n$, there exists a matrix $Q \in \mathbb{R}^{m \times n}$ satisfying $Q^T Q = I_n$ and an upper triangular matrix $R \in \mathbb{R}^{n \times n}$ such that $A = QR$. Further, if A is full-rank and we require the diagonal elements of R to be all positive, Q and R are uniquely determined.*

The decompositions described in the above propositions are called the *Cholesky decomposition* and *(thin) QR decomposition*, respectively. In this book, we always adopt the unique upper triangular matrix R in Proposition 2.2, whose diagonal elements are all positive, as the R-factor of the QR decomposition of a full-rank matrix.

Proposition 2.3 *For any symmetric matrix $A \in \mathrm{Sym}(n)$, there exist $P \in \mathcal{O}_n$ and a diagonal matrix $\Lambda \in \mathbb{R}^{n \times n}$ such that A can be decomposed into $A = P \Lambda P^T$.*

The decomposition in Proposition 2.3 is called the *eigenvalue decomposition*. In fact, for $i = 1, 2, \ldots, n$, the ith diagonal element and ith column of Λ and P are an eigenvalue and the associated eigenvector of A, respectively. To observe this, we multiply $A = P \Lambda P^T$ by P from the right to obtain $AP = P \Lambda$.

More generally, any matrix can be decomposed into a similar form, as follows.

Proposition 2.4 *Assume that $m \geq n$ for simplicity. Any matrix $A \in \mathbb{R}^{m \times n}$ can be decomposed into $A = U \Sigma V^T$, where $U \in \mathbb{R}^{m \times n}$ and $V \in \mathbb{R}^{n \times n}$ satisfy $U^T U = V^T V = I_n$, and Σ is an $n \times n$ diagonal matrix with nonnegative diagonal elements.*

This decomposition is called the *(thin) singular value decomposition*. For the singular value decomposition $A = U \Sigma V^T$, the diagonal elements of Σ and the column vectors of U and V are called the *singular values* and *left-* and *right-singular vectors*, respectively. When A is square, the ratio of the largest to smallest singular value of A is called the (2-norm) *condition number* of A, which is defined as ∞ if A is singular. More generally, the condition numbers of a square matrix are defined via matrix norms, although we omit further details from this book.

A square matrix $X \in \mathbb{R}^{n \times n}$ satisfying $X^2 = A$ is called a *square root* of $A \in \mathbb{R}^{n \times n}$. If A is symmetric positive semidefinite, for the eigenvalue decomposition $A = P \, \mathrm{diag}(\lambda_1, \lambda_2, \ldots, \lambda_n) P^T$, the matrix $X := P \, \mathrm{diag}\left(\sqrt{\lambda_1}, \sqrt{\lambda_2}, \ldots, \sqrt{\lambda_n}\right) P^T$ satisfies $X^2 = A$. Note that each eigenvalue λ_i of A is nonnegative in this case. This X is also symmetric positive semidefinite and denoted by $\sqrt{A}$.

2.2 Inner Product Spaces and Euclidean Spaces

In principle, the vector spaces considered herein are over the field of real numbers $\mathbb{R}$, although vector spaces can be defined over a general field. We omit a complete definition of a vector space and its many properties, but refer readers to, e.g., Axler (2015). Throughout this book, we discuss finite-dimensional vector spaces.

Let m and n be positive integers. One of the simplest examples of vector spaces is the numerical vector space $\mathbb{R}^n := \{(x_1, x_2, \ldots, x_n)^T \mid x_1, x_2, \ldots, x_n \in \mathbb{R}\}$, where the addition $+$ is defined as the element-wise addition and the scalar product of $a \in \mathbb{R}$

and $x = (x_i) \in \mathbb{R}^n$ is defined as the vector $ax := (ax_i) \in \mathbb{R}^n$. Note that $x \in \mathbb{R}^n$ can be regarded both as a point in $\mathbb{R}^n$ and as a position vector that represents the position of the point in relation to the origin $0 = (0, 0, \ldots, 0)^T \in \mathbb{R}^n$. The set of all $m \times n$ matrices, $\mathbb{R}^{m \times n}$, can be regarded as a vector space in a similar manner.

We focus on vector spaces equipped with inner products.

Definition 2.1 For a vector space V over $\mathbb{R}$, we assume that $\langle x, y \rangle \in \mathbb{R}$ is defined for any $x, y \in V$. Then, $\langle x, y \rangle$ is called an *inner product* if the following conditions hold, where $x, y, z \in V$ and $a \in \mathbb{R}$ are arbitrary:

1. $\langle x + y, z \rangle = \langle x, z \rangle + \langle y, z \rangle$.
2. $\langle ax, y \rangle = a\langle x, y \rangle$.
3. $\langle x, y \rangle = \langle y, x \rangle$.
4. $\langle x, x \rangle \geq 0$. Further, $\langle x, x \rangle = 0$ if and only if $x = 0$.

The map $\langle \cdot, \cdot \rangle : V \times V \to \mathbb{R}$ is also called an *inner product* and V equipped with $\langle \cdot, \cdot \rangle$ is called an *inner product space*.

The last two conditions in Definition 2.1 are closely related to the concept of a symmetric positive definite matrix, as the following examples imply.

Example 2.1 Let $G \in \mathrm{Sym}(n)$ be positive definite and define $\langle \cdot, \cdot \rangle_G$ in $\mathbb{R}^n$ as

$$\langle x, y \rangle_G := x^T G y, \quad x, y \in \mathbb{R}^n. \tag{2.1}$$

Then, $\langle \cdot, \cdot \rangle_G$ is an inner product in $\mathbb{R}^n$. In the particular case of $G = I_n$, we obtain

$$\langle x, y \rangle_{I_n} = x^T y = \sum_{i=1}^{n} x_i y_i, \quad x = (x_i), \ y = (y_i) \in \mathbb{R}^n. \tag{2.2}$$

The inner product $\langle \cdot, \cdot \rangle_{I_n}$ thus obtained is called the *standard inner product*, and we call $\mathbb{R}^n$ equipped with (2.2) the *n-dimensional (standard) Euclidean space*.

Example 2.2 Let $G \in \mathrm{Sym}(m)$ be positive definite and define $\langle \cdot, \cdot \rangle_G$ in $\mathbb{R}^{m \times n}$ as

$$\langle X, Y \rangle_G := \mathrm{tr}(X^T G Y), \quad X, Y \in \mathbb{R}^{m \times n}. \tag{2.3}$$

Then, $\langle \cdot, \cdot \rangle_G$ is proved to be an inner product in $\mathbb{R}^{m \times n}$; thus, $\mathbb{R}^{m \times n}$ equipped with (2.3) is an inner product space. As in $\mathbb{R}^n$, we define the standard inner product in $\mathbb{R}^{m \times n}$ by putting $G = I_m$ in (2.3) as

$$\langle X, Y \rangle_{I_m} = \mathrm{tr}(X^T Y) = \sum_{i=1}^{m} \sum_{j=1}^{n} x_{ij} y_{ij}, \quad X = (x_{ij}), \ Y = (y_{ij}) \in \mathbb{R}^{m \times n}.$$

Let V be an n-dimensional inner product space with an arbitrary inner product $\langle \cdot, \cdot \rangle$. Two vectors $x, y \in V$ are said to be *orthogonal* if $\langle x, y \rangle = 0$. For a subspace

W of V, the *orthogonal complement* of W, denoted by $W^\perp$, is the set of vectors in V orthogonal to any vector in W, i.e., $W^\perp := \{x \in V \mid \langle x, y \rangle = 0 \text{ for all } y \in W\}$.

Definition 2.2 If V is a vector space over $\mathbb{R}$ and a map $\|\cdot\|: V \to \mathbb{R}$ satisfies the following conditions, where $x, y \in V$ and $c \in \mathbb{R}$ are arbitrary, then $\|x\|$ is called the *norm* of $x \in V$ and the vector space V equipped with $\|\cdot\|$ is called a *normed space*.

1. $\|x\| \geq 0$. Further, $\|x\| = 0$ if and only if $x = 0$.
2. $\|cx\| = |c|\|x\|$.
3. $\|x + y\| \leq \|x\| + \|y\|$.

Example 2.3 If a vector space V is endowed with an arbitrary inner product $\langle \cdot, \cdot \rangle$, we can induce the norm $\|\cdot\|$ by defining

$$\|x\| := \sqrt{\langle x, x \rangle}, \quad x \in V. \tag{2.4}$$

Hence, an inner product space can be naturally regarded as a normed space.

Example 2.4 In $\mathbb{R}^n$, the *p-norm* of $x = (x_i) \in \mathbb{R}^n$ for a real number $p \geq 1$, defined by

$$\|x\|_p := (|x_1|^p + |x_2|^p + \cdots + |x_n|^p)^{\frac{1}{p}},$$

is important. In particular, the 2-norm $\|x\|_2 := \sqrt{x_1^2 + x_2^2 + \cdots + x_n^2}$, called the *Euclidean norm*, can also be defined using the standard inner product (2.2), as

$$\|x\|_2 = \sqrt{\langle x, x \rangle_{I_n}} = \sqrt{x^T x}.$$

The 1-norm $\|x\|_1 = |x_1| + |x_2| + \cdots + |x_n|$ is also used. This is because minimizing a function of $x \in \mathbb{R}^n$ under the constraint that $\|x\|_1$ is a sufficiently small positive number tends to yield a sparse solution, i.e., one in which most of the elements are 0. Solution sparsity is an important property in statistics and machine learning (Hastie et al. 2015). Nonsmoothness of the 1-norm leads to nonsmooth optimization (see Sect. 6.3.2).

Let V with an arbitrary inner product $\langle \cdot, \cdot \rangle$ be a normed space, as in Example 2.3. It can be proved that the induced norm (2.4) satisfies the *Cauchy–Schwarz inequality* $|\langle x, y \rangle| \leq \|x\|\|y\|$. We can thus define the *angle* between two nonzero vectors x and y in V as $\theta \in [0, \pi]$, such that

$$\cos \theta = \frac{\langle x, y \rangle}{\|x\|\|y\|} \in [-1, 1]. \tag{2.5}$$

If linearly independent vectors $a_1, a_2, \ldots, a_r \in V$ (r does not exceed the dimension of V) are given, we can always construct $u_1, u_2, \ldots, u_r$ that are orthogonal to each other with respect to the inner product $\langle \cdot, \cdot \rangle$ in V. Specifically, we set $u_1 := a_1$ and

$$u_k := a_k - \sum_{i=1}^{k-1} \frac{\langle a_k, u_i \rangle}{\langle u_i, u_i \rangle} u_i \tag{2.6}$$

for $k = 2, 3, \ldots, r$. Furthermore, if we normalize u_k for $k = 1, 2, \ldots, r$ as

$$q_k := \frac{u_k}{\|u_k\|}, \tag{2.7}$$

then $q_1, q_2, \ldots, q_r$ are *orthonormal* vectors, i.e., they are unit vectors and orthogonal to each other. We call such $\{q_1, q_2, \ldots, q_r\}$ an *orthonormal basis* of the subspace $\text{span}\{a_1, a_2, \ldots, a_r\} := \{\sum_{i=1}^{r} c_i a_i \mid c_1, c_2, \ldots, c_r \in \mathbb{R}\}$. Construction of the orthonormal basis using (2.6) and (2.7) is called the *Gram–Schmidt process*.

Remark 2.1 For a full-rank matrix $A \in \mathbb{R}^{m \times n}$, let $q_i \in \mathbb{R}^m$ be the ith column vector of the matrix Q in Proposition 2.2, which satisfies $Q^T Q = I_n$. Then, the condition $Q^T Q = I_n$ is equivalent to $q_i^T q_j = \delta_{ij}$, i.e., $q_1, q_2, \ldots, q_n$ are orthonormal. Thus, QR decomposition can be performed using the Gram–Schmidt process. Furthermore, because the QR decomposition of A is unique, the columns of Q in $A = QR$ can be always written as (2.6) and (2.7).

2.3 Topological Spaces

In this section, we discuss the concept of topological spaces, which are more abstract than vector spaces. For details of topological spaces, see Mendelson (1990), Munkres (2000).

A topology on a set is established by defining open sets.

Definition 2.3 A family $\mathcal{O}$ of subsets of a nonempty set X is called a *topology* on X if it satisfies the following conditions:

1. $X \in \mathcal{O}$ and $\emptyset \in \mathcal{O}$.
2. If $O_1, O_2, \ldots, O_k \in \mathcal{O}$, then $\bigcap_{i=1}^{k} O_i \in \mathcal{O}$.
3. For an arbitrary index set Λ, if $O_\lambda \in \mathcal{O}$ for all $\lambda \in \Lambda$, then $\bigcup_{\lambda \in \Lambda} O_\lambda \in \mathcal{O}$.

Then, $(X, \mathcal{O})$ is called a *topological space*,[1] and $O \in \mathcal{O}$ is called an *open set* in X.

The second condition in the above definition states that the intersection of a finite number of open sets is an open set, whereas the third condition states that the union of any (possibly infinite) number of open sets is an open set.

It is easy to show that the following examples are indeed topological spaces.

Example 2.5 For a nonempty set X, the family $\mathcal{O} := \{X, \emptyset\}$ is a topology on X. This $\mathcal{O}$ is called the *indiscrete topology* and $(X, \mathcal{O})$ is called an *indiscrete space*.

---

[1] We also say that X is a topological space without mentioning $\mathcal{O}$ if there is no risk of confusion. The same can be applied to other mathematical objects.

Example 2.6 Let $\mathcal{O}$ be the power set 2^X of a nonempty set X, i.e., the family of all subsets of X. This $\mathcal{O}$ is a topology on X and is called the *discrete topology*. Then, $(X, \mathcal{O})$ is called a *discrete space*.

Example 2.7 For a subset U of $\mathbb{R}^n$, let us consider the following condition:

$$\text{For any } a \in U, \text{ there exists } \varepsilon > 0 \text{ such that } \{x \in \mathbb{R}^n \mid \|x - a\|_2 < \varepsilon\} \subset U. \quad (2.8)$$

Let $\mathcal{O}$ be the family of all subsets U satisfying the condition (2.8). Then, $\mathcal{O}$ is shown to be a topology on $\mathbb{R}^n$ called the *Euclidean topology*. Hereafter, we simply call $U \subset \mathbb{R}^n$ an open set in $\mathbb{R}^n$ if U is an open set in $(\mathbb{R}^n, \mathcal{O})$, i.e., if U satisfies (2.8).

A subset of a topological space can also be naturally endowed with a topology.

Definition 2.4 Let $(X, \mathcal{O})$ be a topological space and A be a subset of X and define a family of subsets of $A \subset X$ as $\mathcal{O}_A := \{O \cap A \mid O \in \mathcal{O}\}$. Then, $\mathcal{O}_A$ is a topology on A and $(A, \mathcal{O}_A)$ is a topological space. This topology $\mathcal{O}_A$ is called the *relative topology* and $(A, \mathcal{O}_A)$ is called a *topological subspace* of $(X, \mathcal{O})$.

For general sets U and V, a map $f : U \to V$ is said to be a *surjection* if, for any $y \in V$, there exists $x \in U$ such that $y = f(x)$. We call f an *injection* if $f(x_1) = f(x_2)$ implies $x_1 = x_2$. A map f that is both surjective and injective is called a *bijection*. When f is a bijection, the inverse map $f^{-1} : V \to U$ exists.

Definition 2.5 Let $(X_1, \mathcal{O}_1)$ and $(X_2, \mathcal{O}_2)$ be topological spaces. A map $f : X_1 \to X_2$ is called *continuous* if the *preimage* of any open set U in X_2 by f, defined as $f^{-1}(U) := \{x \in X_1 \mid f(x) \in U\}$, is an open set in X_1. If $f : X_1 \to X_2$ is bijective and both f and its inverse f^{-1} are continuous, f is called a *homeomorphism*. If there exists a homeomorphism $f : X_1 \to X_2$, then X_1 and X_2 are said to be *homeomorphic*.

This definition may be somewhat unintuitive, but the condition that $f : X_1 \to X_2$ is continuous in the sense of Definition 2.5 is equivalent to the following: For any $x \in X_1$ and any open set $O' \in \mathcal{O}_2$ containing $f(x) \in X_2$, there exists an open set $O \in \mathcal{O}_1$ containing x such that $f(O) \subset O'$ holds true, where $f(O) := \{f(x) \mid x \in O\} \subset X_2$. In the case where X_1 and X_2 are Euclidean spaces, this condition coincides with the well-known definition of continuity (see Sect. 2.5).

The following definition of convergence in a topological space also reduces to the ordinary definition when considering a Euclidean space, as shown in Example 2.8.

Definition 2.6 A sequence $\{x_n\}$ in a topological space $(X, \mathcal{O})$ is said to *converge* to a point $x \in X$ if, for any open set $O \in \mathcal{O}$ containing x, there exists $N \in \mathbb{N}$ such that $n \geq N$ implies $x_n \in O$.

Example 2.8 Consider the Euclidean space $\mathbb{R}^n$ with the Euclidean topology $\mathcal{O}$ introduced in Example 2.7. A sequence $\{x_n\}$ in $\mathbb{R}^n$ converges to $x \in \mathbb{R}^n$ in the sense of Definition 2.6 if and only if the following well-known property holds true: For any $\varepsilon > 0$, there exists $N \in \mathbb{N}$ such that $n \geq N$ implies $\|x_n - x\|_2 < \varepsilon$.

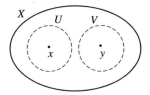

Fig. 2.1 Conceptual illustration of Hausdorff space $(X, \mathcal{O})$. For any points $x, y \in X$, there exist disjoint open neighborhoods $U \in \mathcal{O}$ of x and $V \in \mathcal{O}$ of y, i.e., $x \in U$, $y \in V$, and $U \cap V = \emptyset$

We now define the concept of a Hausdorff space, which is used in the definition of a manifold provided in the next chapter.

Definition 2.7 For a topological space $(X, \mathcal{O})$, if, for any $x, y \in X$ with $x \neq y$, there exist $U, V \in \mathcal{O}$ such that $x \in U$, $y \in V$, and $U \cap V = \emptyset$, then the topology $\mathcal{O}$ is said to satisfy the *Hausdorff separation axiom* and $(X, \mathcal{O})$ is called a *Hausdorff space*.

Figure 2.1 is a conceptual illustration of a Hausdorff space. The Hausdorff separation axiom may seem natural if we simply imagine the situation shown in Fig. 2.1. However, an indiscrete space, which is not a usual space, provides a possibly unintuitive example, which is not a Hausdorff space.

Example 2.9 Suppose that a set X has two or more elements and consider the indiscrete space $(X, \mathcal{O})$, where $\mathcal{O} = \{X, \emptyset\}$. Then, for any $x, y \in X$ with $x \neq y$, $X \in \mathcal{O}$ is the only open set containing x and the only one containing y. Therefore, $(X, \mathcal{O})$ is not a Hausdorff space. Thus, it is natural to call $(X, \{X, \emptyset\})$ an indiscrete space since no two different points can be separated by open sets.

Example 2.10 In contrast, in a discrete space $(X, \mathcal{O})$, two points $x, y \in X$ $(x \neq y)$ can be separated by open sets $\{x\}, \{y\} \in \mathcal{O}$. Thus, a discrete space is a Hausdorff space. It can be shown that $(\mathbb{R}, \mathcal{O})$ in Example 2.7 is also a Hausdorff space.

Hausdorff spaces have the following important property.

Proposition 2.5 *Let $(X, \mathcal{O})$ be a Hausdorff space. If a sequence $\{x_n\}$ in X converges to $x \in X$ and $y \in X$, then $x = y$, i.e., any sequence in a Hausdorff space converges to, at most, one point in X.*

Proof We prove this claim by contradiction. Assume that a sequence $\{x_n\}$ in X converges to distinct $x \in X$ and $y \in X$. Then, from the definition of a Hausdorff space, there exist $U, V \in \mathcal{O}$ such that $x \in U$, $y \in V$, and $U \cap V = \emptyset$. Because $\{x_n\}$ converges to x and y, there exist $N_1, N_2 \in \mathbb{N}$ such that $n \geq N_1$ and $n \geq N_2$ imply $x_n \in U$ and $x_n \in V$, respectively. Therefore, if $n \geq \max\{N_1, N_2\}$, then $x_n \in U$ and $x_n \in V$, which contradicts $U \cap V = \emptyset$. This completes the proof. $\square$

When $(X, \mathcal{O})$ is a Hausdorff space, we call $x \in X$ a *limit* of a sequence $\{x_n\}$ in X and write $\lim_{n \to \infty} x_n = x$ if $\{x_n\}$ converges to x.

The statement of Proposition 2.5 that the limit is unique may seem trivial. However, this property is not necessarily guaranteed in non-Hausdorff spaces, as shown in the following example.

Example 2.11 Let $(X, \mathcal{O})$ be an indiscrete space with two or more elements, which is not a Hausdorff space. Then, any sequence $\{x_n\}$ in $(X, \mathcal{O})$ converges to any point $x \in X$ because x_n belongs to X for all $n \in \mathbb{N}$ and X is the only open set containing x.

Since we expect that a sequence generated by optimization algorithms will not converge to more than one point, we focus on Hausdorff spaces in this book.

Definition 2.8 Let $(X, \mathcal{O})$ be a topological space. A subset $\mathcal{B}$ of $\mathcal{O}$ is called an *open basis* of $\mathcal{O}$ if it satisfies the following condition: For any open set $O \in \mathcal{O}$, there exists $\mathcal{B}_0 \subset \mathcal{B}$ such that $O = \bigcup \{W \mid W \in \mathcal{B}_0\}$.

A topological space that has a "not so large" open basis is an important subject of research. Here, a set S is said to be *countably infinite* if there exists a bijection between S and $\mathbb{N}$. Note that $\mathbb{N}$, which is the set of all natural numbers, is "smaller" than $\mathbb{R}$ in the sense that there is no surjection from $\mathbb{N}$ to $\mathbb{R}$. A set is called a *countable set* if it is either finite or countably infinite. A set that is not countable is called an *uncountable set*.

Definition 2.9 Let $(X, \mathcal{O})$ be a topological space. If there exists an open basis $\mathcal{B}$ of $\mathcal{O}$ such that $\mathcal{B}$ contains a finite or countably infinite number of elements, i.e., open sets in X, then $(X, \mathcal{O})$ is called a *second-countable space*.

We start with an example that is not second-countable and then proceed to second-countable examples.

Example 2.12 Consider an uncountable set X. In the discrete space $(X, \mathcal{O})$ with $\mathcal{O} = 2^X$, for any $x \in X$, any open basis $\mathcal{B}$ of $\mathcal{O}$ must contain $\{x\}$ as its element since $\{x\}$ is open. Therefore, $(X, \mathcal{O})$ is not second-countable.

Example 2.13 An indiscrete space $(X, \mathcal{O})$ with $\mathcal{O} = \{X, \emptyset\}$ is obviously second-countable because $\mathcal{B} := \{X\}$, which consists of only one element of $\mathcal{O}$, is an open basis of $\mathcal{O}$. Note that $\emptyset \in \mathcal{O}$ is the union of no element of $\mathcal{B}$. The Euclidean space $\mathbb{R}^n$ can also be shown to be second-countable (see, e.g., Munkres (2000)).

An indiscrete space with more than one element, which is not a Hausdorff space (Example 2.9), and an uncountable discrete space, which is not second-countable (Example 2.12), are extreme examples. In fact, the Riemannian manifolds considered in this book are assumed to be both Hausdorff spaces and second-countable. The roles of these properties are explained in Chap. 3.

The product set $X_1 \times X_2$ of two topological spaces X_1 and X_2 can be endowed with a natural topology.

Definition 2.10 Let $(X_1, \mathcal{O}_1)$ and $(X_2, \mathcal{O}_2)$ be topological spaces. A family of subsets of $X_1 \times X_2$ defined by $\mathcal{B} := \{O_1 \times O_2 \mid O_1 \in \mathcal{O}_1, \ O_2 \in \mathcal{O}_2\}$ is an open basis of a topology $\mathcal{O}$ of $X_1 \times X_2$. The topology $\mathcal{O}$ and topological space $(X_1 \times X_2, \mathcal{O})$ are called the *product topology* and *product space*, respectively.

From the definition, each open set in the product space $(X_1 \times X_2, \mathcal{O})$ is a union of a finite or infinite number of sets of the form $O_1 \times O_2$ with $O_1 \in \mathcal{O}_1$ and $O_2 \in \mathcal{O}_2$.

The following concept of quotient topology plays an important role in discussions of quotient manifolds.

Definition 2.11 Let $(X, \mathcal{O})$ be a topological space, Y be a set, and $f : X \to Y$ be a surjection, and define a family of subsets of Y as $\mathcal{O}_f := \{O \subset Y \mid f^{-1}(O) \in \mathcal{O}\}$. Then, $\mathcal{O}_f$ is a topology on Y and $(Y, \mathcal{O}_f)$ is a topological space. This topology $\mathcal{O}_f$ is called the *quotient topology* and $(Y, \mathcal{O}_f)$ is called a *quotient space*.

2.4 Groups, Group Actions, and Quotient Spaces

We use the concepts of groups and group actions (see, e.g., Dummit and Foote (2003)) to discuss quotient manifolds.

Definition 2.12 A nonempty set G with a binary operation $G \times G \to G: (a, b) \mapsto ab$ is called a *group* if it satisfies the following properties:

1. $(ab)c = a(bc)$ for any $a, b, c \in G$.
2. There exists $e \in G$ such that $ae = ea = a$ for any $a \in G$, called the *identity element*.
3. For any $a \in G$, there exists an element of G, denoted by a^{-1}, such that $aa^{-1} = a^{-1}a = e$ holds. Here, a^{-1} is called the *inverse element* of a.

Definition 2.13 For a group G and set X, a map $\phi: G \times X \to X$ (resp. $\phi: X \times G \to X$) is called a *left* (resp. *right*) *group action* of G on X if the following hold:

1. $\phi(e, x) = x$ (resp. $\phi(x, e) = x$) for any $x \in X$, where e is the identity element of G.
2. $\phi(g, \phi(h, x)) = \phi(gh, x)$ (resp. $\phi(\phi(x, g), h) = \phi(x, gh)$) for any $g, h \in G$ and $x \in X$.

Definition 2.14 For a set X, a binary relation $\sim$ is called an *equivalence relation* if it satisfies the following conditions, where $x, y, z \in X$ are arbitrary:

1. $x \sim x$.
2. If $x \sim y$, then $y \sim x$.
3. If $x \sim y$ and $y \sim z$, then $x \sim z$.

For any $x \in X$, the set $[x] := \{y \in X \mid x \sim y\}$ is called the *equivalence class* of x. Any element of $[x]$ is called a *representative* of $[x]$. The set of all equivalence classes is called the *quotient set* of X by $\sim$ and denoted by $X/\sim$, i.e., $X/\sim := \{[x] \mid x \in X\}$. The map $\pi : X \to X/\sim$ defined by $\pi(x) := [x]$ is called the *natural projection*.

Note that the natural projection $\pi : X \to X/\sim$ is surjective by definition. If $(X, \mathcal{O})$ is a topological space, the quotient set $X/\sim$ can be naturally endowed with a quotient topology (see also Definition 2.11) as $\mathcal{O}_\pi := \{O \subset X/\sim \mid \pi^{-1}(O) \in \mathcal{O}\}$.

Further, in the situation in Definition 2.13, we can define an equivalence relation $\sim$ by the left (resp. right) group action ϕ as, for $x, y \in X$, $x \sim y$ if and only if there exists $g \in G$ such that $y = \phi(g, x)$ (resp. $y = \phi(x, g)$). For this $\sim$, we denote $X/\sim$ by $G \backslash X$ (resp. X/G). The notation X/G is also sometimes used for the case of left group action instead of $G \backslash X$.

2.5   Maps of Class C^r

In this section, we review the continuity and differentiability of maps from $\mathbb{R}^m$ to $\mathbb{R}^n$. Throughout this section, we let U and V be open sets in $\mathbb{R}^m$ and $\mathbb{R}^n$, respectively, and f be a map from U to V.

As a special case of Definition 2.5, $f : U \to V$ is continuous if, for any $x \in U$ and any open set $O' \subset V$ containing $f(x) \in V$, there exists an open set $O \subset U$ containing x such that $f(O) \subset O'$. Let us fix $x \in U$ and say that f is continuous at x if, for any $\varepsilon > 0$, there exists $\delta > 0$ such that $\|x - y\|_2 < \delta$ implies $\|f(x) - f(y)\|_2 < \varepsilon$. Then, the aforementioned condition of the continuity of $f : U \to V$ is equivalent to the condition that f is continuous at any point in U.

It is well known in calculus that a continuous map $f : U \to V$ can be characterized by its coordinate representation. For any $x \in U$, we have $f(x) \in V \subset \mathbb{R}^n$; hence, we can write $f(x) =: (f_1(x), f_2(x), \ldots, f_n(x))^T$. Then, f is continuous if and only if each $f_i : U \to \mathbb{R}$, $i = 1, 2, \ldots, n$ is continuous. Hence, we can discuss the differentiability of each component f_i to discuss that of $f : U \to V$.

Definition 2.15 Let U and V be open sets in $\mathbb{R}^m$ and $\mathbb{R}^n$, respectively. If all types of partial derivatives of a function $h : U \to \mathbb{R}$ with order of at most r exist and are continuous on U, h is said to be of *class C^r*. If h is of class C^r for any $r \in \mathbb{N}$, then h is said to be of *class C^∞*. We say that $f : U \to V$ is of *class C^r* if each of $f_1, f_2, \ldots, f_n : U \to \mathbb{R}$ is of class C^r, where $f(x) = (f_1(x), f_2(x), \ldots, f_n(x))^T$.

It is also well known that linear combinations of C^r maps $f, g : U \to V$ are also of class C^r. When $V = \mathbb{R}$, the product fg is a C^r function. For the composition $h \circ f$ of $h : V \to \mathbb{R}$ and $f : U \to V$, we have the following important property.

Proposition 2.6 *Let U and V be open sets in $\mathbb{R}^m$ and $\mathbb{R}^n$, respectively. Assume that $f : U \to V$ and $h : V \to \mathbb{R}$ are of class C^1. Denoting $f(x) = (f_1(x), f_2(x), \ldots, f_n(x))^T$ for $x = (x_1, x_2, \ldots, x_m)^T \in U$ and $h(y)$ for $y = (y_1, y_2, \ldots, y_n)^T \in V$, we have*

$$\frac{\partial (h \circ f)}{\partial x_i}(x) = \sum_{j=1}^{n} \frac{\partial h}{\partial y_j}(f(x)) \frac{\partial f_j}{\partial x_i}(x). \tag{2.9}$$

The formula (2.9) is itself important in many practical calculations. Furthermore, it yields the following: For open sets $U \subset \mathbb{R}^m$, $V \subset \mathbb{R}^n$, and $W \subset \mathbb{R}^l$, and C^r maps $f : U \to V$ and $g : V \to W$, the composition $g \circ f : U \to W$ is of class C^r.

The following proposition guarantees that the Hessian matrix of any C^2 function is symmetric (see Sect. 2.7).

Proposition 2.7 *Let U be an open set in $\mathbb{R}^m$. For a C^2 map $f : U \to \mathbb{R}$ and any $i, j \in \{1, 2, \ldots, m\}$, we have*

$$\frac{\partial}{\partial x_i} \left(\frac{\partial f}{\partial x_j} \right) = \frac{\partial}{\partial x_j} \left(\frac{\partial f}{\partial x_i} \right).$$

2.6 Gradients of Functions in Euclidean Spaces

Let us identify the function $f(x_1, x_2, \ldots, x_n)$ of n variables with the function $f(x)$, the variable of which is an n-dimensional vector, where $x = (x_1, x_2, \ldots, x_n)^T$. For simplicity, in the remainder of this chapter, we assume that the domain of f is the whole $\mathbb{R}^n$.

Definition 2.16 The *(Euclidean) gradient*[2] $\nabla f(x) \in \mathbb{R}^n$ of a differentiable function $f : \mathbb{R}^n \to \mathbb{R}$ at $x \in \mathbb{R}^n$ is defined by

$$\nabla f(x) := \left(\frac{\partial f}{\partial x_1}(x), \frac{\partial f}{\partial x_2}(x), \ldots, \frac{\partial f}{\partial x_n}(x) \right)^T. \tag{2.10}$$

The gradient $\nabla f(x)$ is the direction corresponding to the maximum increase in f in some sense. Here, let us observe an equivalent claim: $-\nabla f(x)$ at $x \in \mathbb{R}^n$ is the direction corresponding to the maximum decrease in the function value, which is a generalization of the discussion in Sect. 1.1.1. From Proposition 2.6, differentiating f at $x = (x_i)$ in the direction of $d = (d_i) \in \mathbb{R}^n$ yields

$$\mathrm{D}f(x)[d] := \frac{d}{dt} f(x + td) \bigg|_{t=0} = \sum_{j=1}^{n} \frac{\partial f}{\partial x_j}(x) \cdot \frac{d}{dt}(x_j + td_j) \bigg|_{t=0} = \nabla f(x)^T d, \tag{2.11}$$

where $\mathrm{D}f(x)[d]$ is called the *directional derivative* of f at x in the direction of d. If $\nabla f(x) = 0$, x is called a *critical point* (or *stationary point*) of f, and $\nabla f(x)^T d = 0$ for any d. Here, we fix x and assume that $\nabla f(x) \neq 0$. If $\nabla f(x)^T d \neq 0$, this quantity can take any value, with d being multiplied by some real value. Therefore, we also fix the Euclidean norm of d as 1 to observe how the direction of d affects the value of (2.11). Letting $\theta \in [0, \pi]$ be the angle between $\nabla f(x)$ and d, we obtain $\nabla f(x)^T d = \|\nabla f(x)\|_2 \|d\|_2 \cos \theta = \|\nabla f(x)\|_2 \cos \theta$, which is minimized when $\theta = \pi$, i.e.,

$$d = -\frac{\nabla f(x)}{\|\nabla f(x)\|_2}. \tag{2.12}$$

---

[2] In Riemannian geometry, the symbol ∇ is usually used to denote a linear connection. However, because we do not handle linear connections in this book, we use ∇ only to express Euclidean gradients. In contrast, we denote the Riemannian gradient of a function f defined on a Riemannian manifold as grad f (see Sect. 3.3).

Therefore, $-\nabla f(x)$ is called the *steepest descent direction* of f at x.

However, the claim in the previous paragraph is strongly dependent on the choice of norm. Let the 1-norm of d, instead of the 2-norm, be fixed to 1. Assume that $g := \nabla f(x) \neq 0$ and let g_i denote the ith element of g. For simplicity, we further assume that there is a unique element g_j of g whose absolute value is the largest among all the elements. It then follows from the assumption $\|d\|_1 = 1$ that

$$\nabla f(x)^T d \geq -|g^T d| \geq -\sum_{i=1}^{n} |g_i||d_i| \geq -|g_j| \sum_{i=1}^{n} |d_i| = -|g_j| \|d\|_1 = -|g_j|.$$

The equalities hold true if $d = -\operatorname{sgn}(g_j)e_j$, where $\operatorname{sgn}(g_j) = 1$ if $g_j > 0$ and $\operatorname{sgn}(g_j) = -1$ if $g_j < 0$ (note that $g_j \neq 0$ from the assumption), and e_j is an n-dimensional vector whose jth element is 1 and the other elements are 0. Hence, the steepest descent direction of f in the normed space $\mathbb{R}^n$ equipped with the 1-norm can be considered $-\operatorname{sgn}(g_j)e_j$, instead of the negative gradient $-\nabla f(x)$ mentioned in the previous paragraph. The optimization method in which this direction is taken as a search direction is called the *coordinate descent method* (see, e.g., Boyd and Vandenberghe (2014)).

Considering the above, another steepest descent direction may be obtained if we consider the norm defined by a general inner product in $\mathbb{R}^n$. Let $G \in \operatorname{Sym}(n)$ be positive definite, define the inner product by (2.1) in Example 2.1, and consider the induced norm $\|\cdot\|_G$. Because we have $\left\|\sqrt{G}d\right\|_2^2 = d^T G d = \|d\|_G^2$, under the condition $\|d\|_G = 1$ and the assumption $\nabla f(x) \neq 0$, a discussion similar to (2.12) implies that the d that minimizes $\nabla f(x)^T d = \left(\sqrt{G}^{-1}\nabla f(x)\right)^T \left(\sqrt{G}d\right)$ satisfies $\sqrt{G}d = -\alpha \sqrt{G}^{-1} \nabla f(x)$, where $\alpha > 0$ is a real value for normalization. It follows that

$$d = -\frac{G^{-1}\nabla f(x)}{\|G^{-1}\nabla f(x)\|_G},$$

where the direction $G^{-1}\nabla f(x)$ satisfies $\langle G^{-1}\nabla f(x), v\rangle_G = \nabla f(x)^T v = \mathrm{D}f(x)[v]$ for any $v \in \mathbb{R}^n$. We can conclude that, in an inner product space $\mathbb{R}^n$ endowed with a general inner product $\langle \cdot, \cdot \rangle$, the vector grad $f(x) \in \mathbb{R}^n$ such that

$$\mathrm{D}f(x)[v] = \langle \operatorname{grad} f(x), v\rangle, \quad v \in \mathbb{R}^n, \tag{2.13}$$

instead of (2.11), should be an appropriate "gradient" of f. Then, (2.11) is just a particular case of (2.13) with $\langle \cdot, \cdot \rangle$ being the standard inner product. Based on this idea, the concept of a Riemannian gradient, which is the gradient of a function defined on a Riemannian manifold, is discussed in Sect. 3.3.

The standard inner product and Euclidean norm are often considered in optimization in $\mathbb{R}^n$, and various algorithms have been developed based on the steepest descent direction $-\nabla f(x)$. However, other norms have also been considered for optimization algorithm developments. For example, see Boyd and Vandenberghe (2014), Carlson et al. (2015), Nesterov (2012).

Remark 2.2 When a differentiable function $f: \mathbb{R}^n \to \mathbb{R}$ is specified, the Euclidean gradient $\nabla f(x)$ can be obtained by computing every partial derivative of f. However, it is sometimes easier to use the relation (2.11). In other words, if we compute $\frac{d}{dt} f(x + tv)|_{t=0}$ for arbitrary $v \in \mathbb{R}^n$ and write it in the form $g_x^T v$ with $g_x \in \mathbb{R}^n$, we can conclude that $\nabla f(x) = g_x$, as shown in the following example.

Example 2.14 For the linear function $f(x) := c^T x$ with a constant vector $c \in \mathbb{R}^n$, we have

$$\frac{d}{dt} f(x + tv)\bigg|_{t=0} = \frac{d}{dt}(c^T x + (c^T v)t)\bigg|_{t=0} = c^T v.$$

Therefore, $\nabla f(x) = c$ holds.

For $g(x) := x^T A x$ with a constant square matrix $A \in \mathbb{R}^{n \times n}$, it holds that

$$\frac{d}{dt} g(x + tv)\bigg|_{t=0} = \frac{d}{dt}(x^T A x + (x^T A v + v^T A x)t + (v^T A v)t^2)\bigg|_{t=0}$$
$$= ((A + A^T)x)^T v.$$

Thus, we obtain $\nabla g(x) = (A + A^T)x$. If A is symmetric, it follows that $\nabla g(x) = 2Ax$.

Remark 2.3 If f is a differentiable real-valued function defined on the matrix space $\mathbb{R}^{m \times n}$, we identify $\mathbb{R}^{m \times n}$ with $\mathbb{R}^{mn}$ and define the gradient of f as discussed. In other words, in such a case, $\nabla f(X)$ for $X = (x_{ij}) \in \mathbb{R}^{m \times n}$ is an $m \times n$ matrix defined as

$$\nabla f(X) := \left(\frac{\partial f}{\partial x_{ij}}(X) \right).$$

Then, the directional derivative with respect to $V = (v_{ij}) \in \mathbb{R}^{m \times n}$ satisfies

$$\frac{d}{dt} f(X + tV)\bigg|_{t=0} = \sum_{i=1}^{m} \sum_{j=1}^{n} \frac{\partial f}{\partial x_{ij}}(X) \cdot \frac{d}{dt}(x_{ij} + tv_{ij})\bigg|_{t=0} = \operatorname{tr}(\nabla f(X)^T V).$$

Therefore, as stated in Remark 2.2, if we can compute $\frac{d}{dt} f(X + tV)|_{t=0}$ to have the form $\operatorname{tr}(G_X^T V)$ with $G_X \in \mathbb{R}^{m \times n}$, we obtain $\nabla f(X) = G_X$.

Example 2.15 Let $A \in \mathbb{R}^{m \times m}$ be a constant matrix and define a function f on $\mathbb{R}^{m \times n}$ as $f(X) := \operatorname{tr}(X^T A X)$. Then,

$$\frac{d}{dt} f(X + tV)\bigg|_{t=0} = \operatorname{tr}(X^T A V) + \operatorname{tr}(V^T A X) = \operatorname{tr}(((A + A^T)X)^T V).$$

Therefore, we have $\nabla f(X) = (A + A^T)X$, which is equal to $2AX$ if A is symmetric.

2.7   Hessian Matrices of Functions and Taylor's Theorem

The Hessian matrix of a function $f : \mathbb{R}^n \to \mathbb{R}$, which is closely related to the second-order approximation of f, is defined as follows.

Definition 2.17 The *Hessian matrix* $\nabla^2 f(x)$ of a C^2 function $f : \mathbb{R}^n \to \mathbb{R}$ at $x \in \mathbb{R}^n$ is defined as an $n \times n$ matrix whose (i, j) element is $\partial^2 f(x)/\partial x_i \partial x_j$, i.e.,

$$\nabla^2 f(x) := \left(\frac{\partial^2 f}{\partial x_i \partial x_j}(x) \right).$$

Note that $\nabla^2 f(x)$ of a C^2 function f is a symmetric matrix, owing to Proposition 2.7.

Remark 2.4 Let $f : \mathbb{R}^n \to \mathbb{R}$ be of class C^2. Then, for any $d \in \mathbb{R}^n$, we have

$$\frac{d}{dt} \nabla f(x + td) \bigg|_{t=0} = \left(\frac{d}{dt} \left(\frac{\partial f}{\partial x_i}(x + td) \right) \bigg|_{t=0} \right)$$
$$= \left(\sum_{j=1}^{n} \frac{\partial^2 f}{\partial x_j \partial x_i}(x) d_j \right) = \nabla^2 f(x) d.$$

Therefore, if we can compute $\frac{d}{dt} \nabla f(x + td) \big|_{t=0}$ to write it in the form $H_x d$ with $H_x \in \mathrm{Sym}(n)$, we have $\nabla^2 f(x) = H_x$, as d is arbitrary.

Example 2.16 In Example 2.14, we have $\nabla f(x) = c$ and $\nabla g(x) = (A + A^T)x$, so that $\frac{d}{dt} \nabla f(x + td) \big|_{t=0} = 0$ and $\frac{d}{dt} \nabla g(x + td) \big|_{t=0} = (A + A^T)d$, which yields $\nabla^2 f(x) = 0$ and $\nabla^2 g(x) = A + A^T$. If A is symmetric, this implies $\nabla^2 g(x) = 2A$.

Taylor's theorem is a powerful tool in calculus. Here, we review special cases of Taylor's theorem for a real-valued function $f : \mathbb{R}^n \to \mathbb{R}$ of n variables.

Theorem 2.1 (Taylor's theorem) *Let f be a function from $\mathbb{R}^n$ to $\mathbb{R}$, with x and d belonging to $\mathbb{R}^n$. If f is of class C^1, there exists $\theta_1 \in (0, 1)$ such that*

$$f(x + d) = f(x) + \nabla f(x + \theta_1 d)^T d. \tag{2.14}$$

If f is of class C^2, there exists $\theta_2 \in (0, 1)$ such that

$$f(x + d) = f(x) + \nabla f(x)^T d + \frac{1}{2} d^T \nabla^2 f(x + \theta_2 d) d. \tag{2.15}$$

We know that (2.14) and (2.15) are generalized to the case where f is of class C^r for any r. However, the cases of $r = 1$ and $r = 2$ are particularly important in this book.

Informally, (2.15) states that $f(x + d)$ can be approximately expanded as

$$f(x + d) \approx f(x) + \nabla f(x)^T d + \frac{1}{2} d^T \nabla^2 f(x) d$$

if the norm of d is sufficiently small.

2.8 General Theory of Unconstrained Euclidean Optimization

In this section, we consider the following *unconstrained optimization problem*.

Problem 2.1

$$\begin{aligned} \text{minimize} \quad & f(x) \\ \text{subject to} \quad & x \in \mathbb{R}^n. \end{aligned}$$

This is the problem of finding x that minimizes $f(x)$ subject to the condition $x \in \mathbb{R}^n$, i.e., x can take any vector in $\mathbb{R}^n$. Therefore, we call Problem 2.1 an unconstrained optimization problem in $\mathbb{R}^n$. In the remainder of this chapter, we consider the Euclidean space $\mathbb{R}^n$ equipped with the standard inner product (2.2).

Definition 2.18 A point $x_* \in \mathbb{R}^n$ is called a *global optimal solution* to Problem 2.1 if $f(x_*) \leq f(x)$ holds true for any $x \in \mathbb{R}^n$. A point x_* is called a *local optimal solution*[3] to Problem 2.1 if there exists a neighborhood[4] U of $x_* \in \mathbb{R}^n$ such that $x \in U$ implies that $f(x_*) \leq f(x)$. If there exists a neighborhood U of $x_* \in \mathbb{R}^n$ such that $f(x_*) < f(x)$ holds true for all $x \in U$ except x_*, then x_* is called a *strict local optimal solution*.

2.8.1 Convexity of Functions

Next, we define the convexity of real-valued functions.

Definition 2.19 A function $f : \mathbb{R}^n \to \mathbb{R}$ is called a *convex function* if

$$f(\alpha x + (1 - \alpha)y) \leq \alpha f(x) + (1 - \alpha)f(y) \tag{2.16}$$

holds for any $x, y \in \mathbb{R}^n$ and $\alpha \in [0, 1]$. Furthermore, if f satisfies

$$f(\alpha x + (1 - \alpha)y) < \alpha f(x) + (1 - \alpha)f(y)$$

for any distinct $x, y \in \mathbb{R}^n$ and $\alpha \in (0, 1)$, then f is called a *strictly convex function*.

---

[3] By definition, a global optimal solution is always a local optimal solution.

[4] We always consider open neighborhoods, i.e., a neighborhood of x is an open set containing x.

Example 2.17 Any linear function $f(x) := c^T x$ with constant $c \in \mathbb{R}^n$ is convex, as the equals sign in (2.16) holds true for any $x, y \in \mathbb{R}^n$ and $\alpha \in [0, 1]$.

Let $A \in \text{Sym}(n)$ be positive definite. Then, the quadratic function $g(x) := x^T A x$ is strictly convex because a straightforward computation yields

$$\alpha g(x) + (1 - \alpha)g(y) - g(\alpha x + (1 - \alpha)y) = \alpha(1 - \alpha)(x - y)^T A(x - y) > 0$$

for any $\alpha \in (0, 1)$ and distinct $x, y \in \mathbb{R}^n$. Similarly, if A is symmetric positive semidefinite, $g(x) = x^T A x$ can be proved to be convex.

The following proposition clearly follows from Definition 2.19.

Proposition 2.8 *If $f : \mathbb{R}^n \to \mathbb{R}$ and $g : \mathbb{R}^n \to \mathbb{R}$ are convex functions, then $f + g$ is convex. If we further assume that at least one of f or g is strictly convex, so is $f + g$.*

Convex and strictly convex functions have the following important properties.

Proposition 2.9 *If f is convex and x_* is a local optimal solution to Problem 2.1, then x_* is a global optimal solution.*

Proof By contradiction, assume that there exists $x \in \mathbb{R}^n$ such that $f(x) < f(x_*)$. Then, for any neighborhood U of x_*, $\alpha x + (1 - \alpha)x_*$ with sufficiently small $\alpha \in (0, 1]$ is in U. It follows from the convexity of f that $\alpha x + (1 - \alpha)x_* \in U$ satisfies

$$f(\alpha x + (1 - \alpha)x_*) \leq \alpha f(x) + (1 - \alpha)f(x_*) < \alpha f(x_*) + (1 - \alpha)f(x_*) = f(x_*),$$

contrary to the assumption that x_* is a local optimal solution. This ends the proof. □

Proposition 2.10 *If f is strictly convex and x_* is a local optimal solution to Problem 2.1, then x_* is the unique global optimal solution.*

Proof Because f is a convex function, from Proposition 2.9, x_* is a global optimal solution. To verify that it is indeed a unique global optimal solution by contradiction, we assume that there exists another global optimal solution $x \neq x_*$. Then, it follows from the strict convexity of f and the assumption $f(x) = f(x_*)$ that

$$f\left(\frac{x_* + x}{2}\right) < \frac{1}{2}f(x_*) + \frac{1}{2}f(x) = f(x_*),$$

contradicting the fact that x_* is a global optimal solution. This ends the proof. □

There are other important properties of convex functions. If f is of class C^2, f is convex if and only if $\nabla^2 f(x)$ is positive semidefinite for all $x \in \mathbb{R}^n$. Furthermore, if $\nabla^2 f(x)$ is positive definite for all $x \in \mathbb{R}^n$, then f is strictly convex, but not vice versa (consider the strictly convex function $f(x) := x^4$ on $\mathbb{R}$ and $x = 0$). Readers may refer to Bertsekas (2016), Nocedal and Wright (2006) for further details.

2.8.2 Optimality Conditions

The gradient and Hessian matrix of the objective function f help us check if a given $x \in \mathbb{R}^n$ is a local optimal solution to Problem 2.1.

Theorem 2.2 *If f is of class C^1 and $x_* \in \mathbb{R}^n$ is a local optimal solution to Problem 2.1, then we have $\nabla f(x_*) = 0$.*

Proof To prove the theorem by contradiction, we assume $\nabla f(x_*) \neq 0$. Let d be $-\nabla f(x_*)$. Because $\nabla f(x_*)^T d = -\|\nabla f(x_*)\|_2^2 < 0$ and ∇f is continuous, there exists $\bar{s} > 0$ such that $\nabla f(x_* + sd)^T d < 0$ for any $s \in [0, \bar{s}]$. For any nonzero $s \in (0, \bar{s}]$, it follows from Taylor's theorem that there exists $t \in (0, s)$ such that

$$f(x_* + sd) = f(x_*) + s\nabla f(x_* + td)^T d.$$

Because $t < s \leq \bar{s}$, we have $\nabla f(x_* + td)^T d < 0$ and conclude that $f(x_* + sd) < f(x_*)$. Here, s can be an arbitrarily small positive number. This contradicts the fact that x_* is a local optimal solution, thereby completing the proof. □

Theorem 2.3 *If f is of class C^2 and $x_* \in \mathbb{R}^n$ is a local optimal solution to Problem 2.1, then $\nabla f(x_*) = 0$ and $\nabla^2 f(x_*)$ is positive semidefinite.*

Proof Again, we prove the claim by contradiction, i.e., we assume that $\nabla^2 f(x_*)$ is not positive semidefinite. Then, there exists $d \in \mathbb{R}^n - \{0\}$ such that $d^T \nabla^2 f(x_*)d < 0$. The continuity of $\nabla^2 f$ implies that there exists $\bar{s} > 0$ such that $d^T \nabla^2 f(x_* + sd)d < 0$ for any $s \in [0, \bar{s}]$. For any $s \in (0, \bar{s}]$, from Taylor's theorem, there exists $t \in (0, s)$ such that

$$f(x_* + sd) = f(x_*) + s\nabla f(x_*)^T d + \frac{1}{2}s^2 d^T \nabla^2 f(x_* + td)d.$$

Theorem 2.2 yields $\nabla f(x_*) = 0$, which, together with $d^T \nabla^2 f(x_* + td)d < 0$ from $t < \bar{s}$, gives $f(x_* + sd) < f(x_*)$. Similar to the proof of Theorem 2.2, this contradicts the local optimality of x_*, thereby completing the proof. □

The conditions on $\nabla f(x_*)$ and $\nabla^2 f(x_*)$ for a local optimal solution x_* described in Theorems 2.2 and 2.3 are called the *first-* and *second-order necessary optimality conditions*, respectively.

Unfortunately, even when $\nabla f(x) = 0$ holds and $\nabla^2 f(x)$ is positive semidefinite for some $x \in \mathbb{R}^n$, such an x is not necessarily guaranteed to be a local optimal solution. That is, the necessary optimality conditions are not sufficient conditions for local optimality (consider, e.g., $f(x_1, x_2) := x_1^4 - x_2^4$ and $(x_1, x_2) = (0, 0)$).

In fact, if we consider the positive definiteness of the Hessian at the point in question, we obtain the *second-order sufficient optimality conditions* as follows.

Theorem 2.4 *If $\nabla f(x_*) = 0$ and $\nabla^2 f(x_*)$ is positive definite for some $x_* \in \mathbb{R}^n$, then x_* is a strict local optimal solution to Problem 2.1.*

Proof As $\nabla^2 f(x_*)$ is positive definite, any $d \in \mathbb{R}^n - \{0\}$ satisfies $d^T \nabla^2 f(x_*)d > 0$. Further, as $\nabla^2 f$ is continuous, there exists $\bar{s} > 0$ such that $d^T \nabla^2 f(x_* + sd)d > 0$ for any $s \in [0, \bar{s}]$. For arbitrarily small $s \in (0, \bar{s}]$, it follows from the condition $\nabla f(x_*) = 0$ and Taylor's theorem that there exists $t \in (0, s)$ such that

$$f(x_* + sd) = f(x_*) + s\nabla f(x_*)^T d + \frac{1}{2}s^2 d^T \nabla^2 f(x_* + td)d > f(x_*).$$

Because $s > 0$ can be taken as arbitrarily small and d is an arbitrary direction, there exists a sufficiently small neighborhood U of x_* such that $f(x)$ at any $x \in U - \{x_*\}$ is larger than $f(x_*)$, i.e., x_* is a strict local optimal solution. □

2.9 Unconstrained Optimization Methods in Euclidean Spaces

This section provides some approaches to numerically solving unconstrained optimization problems. Throughout this section, we suppose that the objective function f of Problem 2.1 is of class C^1.

2.9.1 Line Search Method Framework and Step Lengths

In this subsection, we introduce a framework for the optimization methods called *line search methods*. In a line search method, we start from a given initial point $x_0 \in \mathbb{R}^n$ to generate a sequence $\{x_k\}$ in $\mathbb{R}^n$ using the updating formula

$$x_{k+1} = x_k + t_k d_k, \quad k = 0, 1, \ldots, \tag{2.17}$$

which is expected to converge to an optimal solution. Here, $d_k \in \mathbb{R}^n$ and $t_k > 0$ are called a *search direction* and *step length*, respectively, the choices of which characterize many optimization algorithms. We usually expect that the value of the objective function decreases along the search direction, i.e., $f(x_k + t_k d_k)$ with a sufficiently small step length t_k is less than $f(x_k)$. Hence, we consider the condition

$$\lim_{t \to +0} \frac{f(x_k + td_k) - f(x_k)}{t} < 0.$$

Because the left-hand side equals $Df(x_k)[d_k] = \nabla f(x_k)^T d_k$, the condition is rewritten as

$$\nabla f(x_k)^T d_k < 0. \tag{2.18}$$

A search direction d_k satisfying (2.18) is called a *descent direction* of f at x_k.

Fig. 2.2 Line search method for Problem 2.1 in $\mathbb{R}^n$. The figure illustrates generation of a sequence $\{x_k\} \subset \mathbb{R}^n$ for $n = 2$. For each $k \geq 0$, from the current point x_k, we compute the next point x_{k+1} as $x_{k+1} = x_k + t_k d_k$ with a search direction d_k and step length t_k

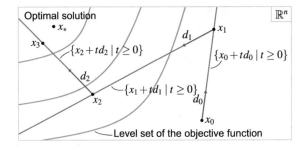

Algorithm 2.1 constitutes a general line search method framework. The algorithm concept is illustrated in Fig. 2.2.

Algorithm 2.1 General line search method for Problem 2.1.

Input: Objective function f and initial point $x_0 \in \mathbb{R}^n$.
Output: Sequence $\{x_k\} \subset \mathbb{R}^n$.
1: **for** $k = 0, 1, 2, \ldots$ **do**
2:    Compute a search direction $d_k \in \mathbb{R}^n$.
3:    Compute a step length $t_k > 0$.
4:    Compute the next point as $x_{k+1} = x_k + t_k d_k$.
5: **end for**

Remark 2.5 In addition to line search methods, trust-region methods are also well known, although they are not discussed herein. See Sect. 6.3.1 for references.

We next discuss determination of t_k in (2.17) when $x_k, d_k \in \mathbb{R}^n$ are already computed at the kth iteration. Finding an appropriate $t_k > 0$ is equivalent to finding an appropriate next point on the half-line $\{x_k + t d_k \mid t > 0\}$ emanating from x_k in the direction of d_k, and is closely related to minimization of the one variable function

$$\phi_k(t) := f(x_k + t d_k) \tag{2.19}$$

under the condition $t > 0$. Thus, it seems theoretically natural to determine t_k as

$$t_k := \arg\min_{t>0} \phi_k(t) = \arg\min_{t>0} f(x_k + t d_k) \tag{2.20}$$

if a unique minimizer exists. Selection of t_k by (2.20) is called *exact line search*.

Example 2.18 Consider the quadratic objective function $f(x) := \frac{1}{2} x^T A x - b^T x$, where $A \in \text{Sym}(n)$ is positive definite and $b \in \mathbb{R}^n$. If we assume that the search direction d_k is a descent direction, i.e., that (2.18) holds, then $d_k \neq 0$ and

$$f(x_k + t d_k) = \frac{1}{2}(d_k^T A d_k)t^2 + (\nabla f(x_k)^T d_k)t + f(x_k),$$

Fig. 2.3 Graphs of curve
$C : y = \phi_k(t)$, tangent line
$y = \phi_k'(0)t + \phi_k(0)$, and line
$l : y = c_1\phi_k'(0)t + \phi_k(0)$
and intervals in which (2.22)
is satisfied. Any t in the
intervals can be acceptable
when we use the Armijo
condition

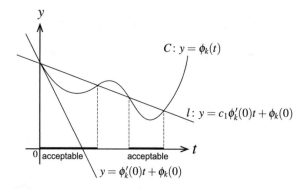

where $\nabla f(x) = Ax - b$. Because the positive definiteness of A and $d_k \neq 0$ yield
$d_k^T A d_k > 0$, the above quantity is minimized when

$$t = -\frac{\nabla f(x_k)^T d_k}{d_k^T A d_k}, \tag{2.21}$$

which is positive from (2.18). Thus, t_k through exact line search (2.20) is given
by (2.21). We observed these aspects using a simple example in Sect. 1.1.1.

It may not be easy to compute t_k through exact line search for a general, not
necessarily quadratic, function. Hence, t_k is chosen such that it satisfies certain conditions, which guarantee that the selected t_k is "good" to some extent. To discuss
these criteria, it is convenient to use ϕ_k defined by (2.19).

Consider a graph $C : y = \phi_k(t)$ on the t–y plane, as in Fig. 2.3. If d_k is a descent
direction at x_k, i.e., (2.18) holds, then $\phi_k'(0) = \frac{d}{dt} f(x_k + t d_k)\big|_{t=0} = \nabla f(x_k)^T d_k <$
0. Therefore, the slope of the tangent of the graph C at $(0, \phi_k(0))$ is negative. Let
c_1 be a constant with $0 < c_1 < 1$ and consider the line $l : y = c_1\phi_k'(0)t + \phi_k(0)$,
whose slope is gentler than that of the tangent of C at $(0, \phi_k(0))$. Let us choose a
step length t_k such that the curve C is below the line l or intersects with l at $t = t_k$;
i.e.,

$$\phi_k(t_k) \leq c_1\phi_k'(0)t_k + \phi_k(0), \tag{2.22}$$

which is rewritten from the definition of ϕ_k as

$$f(x_k + t_k d_k) \leq f(x_k) + c_1 t_k \nabla f(x_k)^T d_k. \tag{2.23}$$

Inequality (2.23) is called the *Armijo condition*. Figure 2.3 illustrates the values of t
acceptable for t_k with respect to the Armijo condition.

If d_k is a descent direction, a sufficiently small t_k satisfies the Armijo condition.
Therefore, even if we choose t_k satisfying the Armijo condition at each iteration, x_k
and x_{k+1} are almost identical when t_k is extremely small. To avoid such excessively
small t_k, we can further use the information of $\phi_k'(t)$. Recall that the step length t_k^*

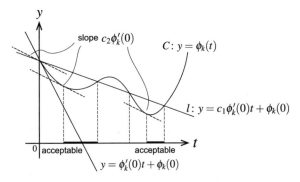

Fig. 2.4 Intervals in which Wolfe conditions (2.22) and (2.24) are satisfied. In addition to Fig. 2.3, dashed lines with slope $c_2\phi'_k(0)$ are shown for information. Compared with Fig. 2.3, the values of t at which the slope $\phi'_k(t)$ of the tangent line of C is less than $c_2\phi'_k(0)$ are excluded

obtained through exact line search satisfies $\phi'_k(t^*_k) = 0$. If t is sufficiently small, $\phi'_k(t)$ is nearly equal to $\phi'_k(0) < 0$; hence, it may be possible to make the slope of the tangent at $(t, \phi(t))$ gentler, i.e., to increase $\phi'_k(t)$, which is negative when t is sufficiently close to 0, by increasing the value of t from 0. Therefore, we take a constant c_2 satisfying $0 < c_1 < c_2 < 1$ and consider the condition

$$\phi'_k(t_k) \geq c_2\phi'_k(0) \tag{2.24}$$

on t_k, which is equivalent to

$$\nabla f(x_k + t_k d_k)^T d_k \geq c_2 \nabla f(x_k)^T d_k. \tag{2.25}$$

Inequalities (2.23) and (2.25) are called the *Wolfe conditions*. Figure 2.4 illustrates the acceptable intervals with respect to the Wolfe conditions. Hence, we can observe that excessively small step lengths are excluded.

Furthermore, noting that (2.24) implies nothing if $\phi'_k(t_k)$ is positive, it is also natural to consider the condition

$$|\phi'_k(t_k)| \leq c_2|\phi'_k(0)|, \tag{2.26}$$

i.e.,

$$|\nabla f(x_k + t_k d_k)^T d_k| \leq c_2|\nabla f(x_k)^T d_k| \tag{2.27}$$

instead of (2.25), to reflect our expectation that $\phi'_k(t_k)$ is close to 0. Inequalities (2.23) and (2.27) are called the *strong Wolfe conditions*.

The strong Wolfe conditions may appear to be the best among the three alternatives. However, the Wolfe conditions are stricter than the Armijo condition and the strong Wolfe conditions are stricter than the Wolfe conditions, and it is more difficult to find a step length satisfying stricter conditions. Therefore, a weaker condition such as the Armijo condition, which is relatively easy to satisfy, is also valuable for possible better performance of the employed optimization method.

2.9.2 Algorithm Convergence

In the previous subsections, we introduced the line search method framework. As discussed above, such methods are iterative, and we expect them to generate a sequence converging to an optimal solution.

Informally, an iterative method is said to globally converge to a (possibly local) optimal solution if the method with any initial point generates a sequence that converges to the solution. However, it may be generally difficult to guarantee such a property. In this book, we say that an iterative method *globally converges* if the sequence $\{x_k\} \subset \mathbb{R}^n$ generated by the method with any initial point $x_0 \in \mathbb{R}^n$ satisfies $\lim_{k \to \infty} \|\nabla f(x_k)\| = 0$ or $\liminf_{k \to \infty} \|\nabla f(x_k)\| = 0$.[5] As discussed in the subsequent subsections, the steepest descent method has a global convergence property whereas Newton's method generally does not.

2.9.3 Steepest Descent and Conjugate Gradient Methods

As explained in Sect. 1.1.1, the *steepest descent method* is one of the most basic optimization algorithms. In this approach, x_{k+1} is computed as $x_{k+1} = x_k - t_k \nabla f(x_k)$, i.e., the search direction is computed as $d_k := -\nabla f(x_k)$ in (2.17). Again, we note that the search direction $-\nabla f(x_k)$ is the steepest descent direction at x_k in $\mathbb{R}^n$ endowed with the standard inner product, as discussed in Sect. 2.6. The algorithm is very simple, but its convergence may be slow. However, since this method is simple, it can be the basis for various other algorithms; thus, it is important.

The *conjugate gradient method* was originally proposed as an iterative method called the *linear conjugate gradient method* for solving linear systems. It was then extended to an algorithm for solving more general unconstrained optimization problems, called the *nonlinear conjugate gradient method*. A detailed description of both methods is provided in Chap. 4. Here, we introduce the (nonlinear) conjugate gradient method as a modified version of the steepest descent method. The initial search direction is just the steepest descent direction; $d_0 := -\nabla f(x_0)$. However, for $k \geq 1$, d_k is taken as the addition of the steepest descent direction and the previous search direction scaled by an appropriately computed real number β_k, i.e., $d_k := -\nabla f(x_k) + \beta_k d_{k-1}$. Numerous options have been proposed for β_k.

The steepest descent and conjugate gradient methods are generalized to those on Riemannian manifolds in Chaps. 3 and 4, respectively.

---

[5] When discussing Riemannian optimization, we replace $\mathbb{R}^n$ with a Riemannian manifold M and ∇f with the Riemannian gradient grad f. See Sect. 3.3 for Riemannian gradients.

2.9.4 Newton and Quasi-Newton Methods

In this subsection, we suppose that the objective function f is of class C^2. Assuming that the Hessian matrix $\nabla^2 f(x_*)$ at a local optimal solution x_* is positive definite and that x_k is sufficiently close to x_* that $\nabla^2 f(x_k)$ is also positive definite, *Newton's method* generates a sequence $\{x_k\}$ in $\mathbb{R}^n$ such that x_{k+1} is computed as a minimizer of a second-order approximation of f at x_k.

We observe that the truncated Taylor expansion of $f(x_k + d)$ at x_k with $d \in \mathbb{R}^n$ up to second order is

$$f(x_k + d) \approx f(x_k) + \nabla f(x_k)^T d + \frac{1}{2} d^T \nabla^2 f(x_k) d =: m_k(d).$$

For the function $m_k(d)$ of variable d, we have $\nabla m_k(d) = \nabla^2 f(x_k)d + \nabla f(x_k)$ and $\nabla^2 m_k(d) = \nabla^2 f(x_k)$ (see Examples 2.14 and 2.16). It follows from the positive definiteness of $\nabla^2 f(x_k)$ and Theorem 2.4 that a critical point of m_k is a strict local optimal solution to the minimization problem of m_k. Furthermore, from Proposition 2.8 and Example 2.17, m_k is strictly convex, which, along with Proposition 2.10, implies that the critical point is indeed the unique global optimal solution.

Let d_k be the vector that minimizes m_k. Then, we have $\nabla m_k(d_k) = 0$, i.e.,

$$\nabla^2 f(x_k)d_k = -\nabla f(x_k). \tag{2.28}$$

Thus, we theoretically conclude that $d_k = -\nabla^2 f(x_k)^{-1} \nabla f(x_k)$. However, instead of computing the inverse matrix $\nabla^2 f(x_k)^{-1}$ and the product with $-\nabla f(x_k)$, the linear system (2.28) should usually be solved for d_k in a numerical computation. Newton's method uses this d_k as its search direction.

In summary, Newton's method theoretically generates a sequence by

$$x_{k+1} = x_k - \nabla^2 f(x_k)^{-1} \nabla f(x_k).$$

Note that, unlike the steepest descent or conjugate gradient methods, Newton's method exploits the Hessian matrix, i.e., the second-order information of f. Additionally, the computational cost of solving the linear system (2.28) is usually the highest in one iteration of Newton's method. Although Newton's method is a fast algorithm that has "a locally quadratic convergence property" under certain conditions, it is unsuitable for use when the dimension of the linear system (2.28) is extremely large because of the resultant long computation time required in one iteration. Furthermore, an initial point for Newton's method should be sufficiently close to an optimal solution. Therefore, Newton's method is useful especially when an adequate approximate solution is obtained through some other method or from the specific structure of the problem to find a more accurate solution.

To avoid solving (2.28), the *quasi-Newton method* approximates the Hessian matrix $\nabla^2 f(x_k)$ or its inverse $\nabla^2 f(x_k)^{-1}$ using a quantity that can be computed relatively easily. Although we omit a detailed derivation of these algorithms, several famous formulas for approximating the Hessian or its inverse are available. One of the most well-known methods is the *BFGS (Broyden–Fletcher–Goldfarb–Shanno) method* (Fletcher 2000; Nocedal and Wright 2006).

2.10 Constrained Euclidean Optimization

Although this book mainly focuses on unconstrained optimization on Riemannian manifolds hereafter, we emphasize that such optimization problems include certain *constrained optimization problems* in Euclidean spaces. For example, in Sect. 1.1.2, we reformulated a constrained Euclidean optimization problem as an equivalent unconstrained problem on the sphere, i.e., a Riemannian manifold.

Constrained problems in Euclidean spaces can be solved using existing methods such as the *augmented Lagrange method* and *sequential quadratic programming (SQP)* (Nocedal and Wright 2006). The connection between SQP and Riemannian optimization is reviewed in, e.g., Mishra and Sepulchre (2016). It is known that some constrained optimization algorithms may suffer from the *Maratos effect*, which prevents rapid convergence because step lengths that can make a good progress are rejected (Nocedal and Wright 2006). If the feasible set forms a Riemannian manifold, unconstrained Riemannian optimization can be one way to avoid the Maratos effect.

See Sect. 6.2 for remarks on constrained Riemannian optimization.

Chapter 3
Unconstrained Optimization on Riemannian Manifolds

In this chapter, we introduce the concept of a Riemannian manifold, based on which Riemannian optimization is developed. We also introduce the concept of a retraction, which is important when searching for the next point in an optimization procedure. Furthermore, using a retraction, we discuss one of the simplest optimization methods, the steepest descent method, on Riemannian manifolds.

3.1 Differentiable Manifolds

It is essential to use information about the derivative of an objective function f defined on a set M to minimize it via methods such as the steepest descent and conjugate gradient methods. If we discuss Euclidean spaces only, we cannot directly handle nonlinear spaces, such as the sphere discussed in Sect. 1.1.2. However, if we only assume that M is a topological space with no additional structure, we cannot define the derivative of f since this assumption is quite abstract. This leads us to the concept of a manifold. See, e.g., Lee (1997) for details of Riemannian geometry.

3.1.1 Differentiable Manifolds and Differentiability of Maps

By adding more structure to a topological space M, we aim to define the derivative of f on M. We start with topological manifolds and proceed to differentiable manifolds.

Definition 3.1 A topological space M is called an n-dimensional *topological manifold* if it is a second-countable Hausdorff space and satisfies the condition that, for any point p on M, there exist an open set U in M containing p and a homeomorphism $\varphi : U \to V$ from U to an open set V in $\mathbb{R}^n$. Then, we write $\dim M = n$.

© Springer Nature Switzerland AG 2021
H. Sato, *Riemannian Optimization and Its Applications*,
SpringerBriefs in Control, Automation and Robotics,
https://doi.org/10.1007/978-3-030-62391-3_3

Fig. 3.1 Charts in atlas that overlap on n-dimensional manifold M. The figure illustrates the case of $n = 2$. A point p that belongs to $U_\alpha \cap U_\beta$ can be represented by the two expressions $\varphi_\alpha(p) \in \mathbb{R}^n$ and $\varphi_\beta(p) \in \mathbb{R}^n$. The transition map $\varphi_\beta \circ \varphi_\alpha^{-1}$ is required to be of class C^r when M is a C^r differentiable manifold

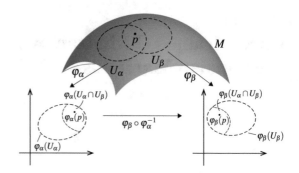

The pair (U, φ) in Definition 3.1 is called a *chart*. In contrast with a general topological space, for the manifold case, if $p \in U$, then $\varphi(p) \in \mathbb{R}^n$ holds, and p can be represented by an n-tuple of real numbers through φ. If $x_i : U \to \mathbb{R}$ maps $p \in U$ to the ith element of $\varphi(p)$, the chart (U, φ) is also denoted by $(U; x_1, x_2, \ldots, x_n)$ and we write $\varphi = (x_1, x_2, \ldots, x_n)$. By definition, for an n-dimensional topological manifold M, there exist a set of indices, Λ, and a family of charts of M, $\{(U_\lambda, \varphi_\lambda)\}_{\lambda \in \Lambda}$, such that $M = \bigcup_{\lambda \in \Lambda} U_\lambda$ holds. The family $\{(U_\lambda, \varphi_\lambda)\}_{\lambda \in \Lambda}$ is called an *atlas* of M.

For any point $p \in M$ and a chart (U, φ) around p, noting that φ is a homeomorphism, U is homeomorphic to an open set $\varphi(U) \subset \mathbb{R}^n$. Hence, one may try to define the differentiability of $f : M \to \mathbb{R}$ at $p \in U$ as that of $f \circ \varphi^{-1} : \varphi(U) \to \mathbb{R}$ at $\varphi(p)$. However, more than one chart of p may exist, and the differentiability of f must be well defined independently of the chart choice. This leads us to the following definition of a differentiable manifold; we impose a condition that the differentiability of $f \circ \varphi^{-1}$ and $f \circ \psi^{-1}$ coincide, where (U, φ) and (V, ψ) are charts around p.

Definition 3.2 Let r be a natural number or ∞. A topological space M is called a C^r *(differentiable) manifold* if M is a topological manifold with an atlas $\{(U_\lambda, \varphi_\lambda)\}_{\lambda \in \Lambda}$ and, for any $\alpha, \beta \in \Lambda$ with $U_\alpha \cap U_\beta \neq \emptyset$, $\varphi_\beta \circ \varphi_\alpha^{-1} : \varphi_\alpha(U_\alpha \cap U_\beta) \to \varphi_\beta(U_\alpha \cap U_\beta)$ is of class C^r. Then, the atlas is called a C^r *atlas*.

For charts $(U_\alpha, \varphi_\alpha)$ and (U_β, φ_β), a point $p \in U_\alpha \cap U_\beta$ is represented by local coordinates $\varphi_\alpha(p) \in \mathbb{R}^n$ via φ_α and $\varphi_\beta(p) \in \mathbb{R}^n$ via φ_β. The above definition indicates that the coordinate transformation $\varphi_\beta \circ \varphi_\alpha^{-1}$ is of class C^r (see Fig. 3.1). Therefore, if $f \circ \varphi_\beta^{-1}$ is of class C^s for some $s \leq r$ at $\varphi_\beta(p)$, then $f \circ \varphi_\alpha^{-1} = (f \circ \varphi_\beta^{-1}) \circ (\varphi_\beta \circ \varphi_\alpha^{-1})$ is also of class C^s at $\varphi_\alpha(p)$.

Example 3.1 We endow the Euclidean space $\mathbb{R}^n$ with the atlas $\mathcal{A} := \{(\mathbb{R}^n, \mathrm{id})\}$ consisting of just one chart. Then, $\mathbb{R}^n$ is clearly an n-dimensional C^∞ manifold.

Example 3.2 The sphere $S^{n-1} := \{x = (x_1, x_2, \ldots, x_n)^T \in \mathbb{R}^n \mid x^T x = 1\}$ can be an $(n - 1)$-dimensional C^∞ manifold. For simplicity, we discuss the case of $n = 3$. We endow S^2 with the relative topology from the Euclidean topology on $\mathbb{R}^3$. It is then obvious that S^2 is a Hausdorff space and is second-countable. As an atlas, we can take $\mathcal{A} := \{(U_1^+, \varphi_1^+), (U_1^-, \varphi_1^-), (U_2^+, \varphi_2^+), (U_2^-, \varphi_2^-), (U_3^+, \varphi_3^+), (U_3^-, \varphi_3^-)\}$, where $U_i^+ := \{x \in S^2 \mid x_i > 0\}$ and $U_i^- := \{x \in S^2 \mid x_i < 0\}$ for $i = 1, 2, 3$, $\varphi_1^\pm(x) := (x_2, x_3)^T$, $\varphi_2^\pm(x) := (x_1, x_3)^T$, and $\varphi_3^\pm(x) := (x_1, x_2)^T$.

Assume that $x \in U_1^+$. Then, we have $\varphi_1^+(x) = (x_2, x_3)^T$. Letting $(y_1, y_2) := (x_2, x_3)$ yields $x_1 = \sqrt{1 - y_1^2 - y_2^2} > 0$. Hence, $(\varphi_1^+)^{-1}(y_1, y_2) = \left(\sqrt{1 - y_1^2 - y_2^2}, y_1, y_2\right)^T$, and $\varphi_1^+ : U_1^+ \to \varphi_1^+(U_1^+)$ is a homeomorphism. The same discussion can be applied to the other charts. Furthermore, the coordinate transformation between each pair of overlapping charts $(U_i^\pm, \varphi_i^\pm)$ and $(U_j^\pm, \varphi_j^\pm)$ is proved to be of class C^∞. For example, for $x \in U_1^+ \cap U_2^+ = \{x \in S^2 \mid x_1 > 0, \ x_2 > 0\}$, we have

$$\varphi_2^+ \circ (\varphi_1^+)^{-1}(y_1, y_2) = \varphi_2^+\left(\sqrt{1 - y_1^2 - y_2^2}, \ y_1, \ y_2\right) = \left(\sqrt{1 - y_1^2 - y_2^2}, \ y_2\right)^T,$$

which is of class C^s for $s \leq r$ because $1 - y_1^2 - y_2^2 > 0$ for $(y_1, y_2)^T \in \varphi_1^+(U_1^+ \cap U_2^+)$. The other cases are also similar. We thus conclude that $\mathcal{A}$ is a C^∞ atlas of S^2.

More generally, S^{n-1} also has a similar atlas and can thus be a C^∞ manifold.

The above example shows a concrete choice of atlas for the sphere. However, as discussed in Sect. 1.1.2, when optimization algorithms are performed numerically, local coordinates are not necessarily used directly. Nevertheless, theoretically, local coordinates may sometimes be fundamental in, e.g., local convergence analysis of optimization algorithms such as Newton's method.

We discuss atlases further. There may exist more than one C^r atlas on M. Assume that M has two different atlases $\mathcal{A} = \{(U_\alpha, \varphi_\alpha)\}_{\alpha \in A}$ and $\mathcal{B} = \{(V_\beta, \psi_\beta)\}_{\beta \in B}$ on M. If $\psi_\beta \circ \varphi_\alpha^{-1}$ and $\varphi_\alpha \circ \psi_\beta^{-1}$ are of class C^r for any $\alpha \in A$ and $\beta \in B$ with $U_\alpha \cap V_\beta \neq \emptyset$, then $\mathcal{A} \cup \mathcal{B}$ is also a C^r atlas of M. In such a case, we say that the two atlases are *equivalent*. It can be proved that, if $f : M \to \mathbb{R}$ is of class C^s for $s \leq r$ (see Definition 3.4) with respect to $\mathcal{A}$, it is also of class C^s with respect to $\mathcal{B}$ that is equivalent to $\mathcal{A}$. Therefore, it is not necessary to adhere to one of the two atlases. Thus, we define the concept of a maximal atlas as follows.

Definition 3.3 Let M be a C^r manifold endowed with an atlas $\mathcal{A}$. The union of all atlases equivalent to $\mathcal{A}$ is called the *maximal atlas* or *differential structure* of M generated by $\mathcal{A}$. Any chart in the maximal atlas is called a C^r *chart*.

We assume a maximal atlas unless we endow a manifold with a specific atlas.

We are now in a position to define the differentiability of functions.

Definition 3.4 Let M be a C^r manifold with $r \geq 0$ and s an integer satisfying $0 \leq s \leq r$. A function $f : M \to \mathbb{R}$ is said to be of class C^s if, for every $p \in M$, there exists a chart (U, φ) such that $f \circ \varphi^{-1}$ is of class C^s at $\varphi(p)$.

To summarize, although f is a function on a manifold M, for each chart (U, φ), it is represented via φ as $f \circ \varphi^{-1}$, which enables the usual differentiation performed in Euclidean spaces. Definition 3.4 exploits the differentiability of $f \circ \varphi^{-1}$ at $\varphi(p)$ to define the differentiability of f at $p \in U$, which does not depend on the chart choice because $\varphi_\beta \circ \varphi_\alpha^{-1}$ in Definition 3.2 is of class C^r.

However, $f \circ \varphi^{-1}$ itself depends on the chart choice. Therefore, we do not define the derivative of f using, e.g., $\partial(f \circ \varphi^{-1})/\partial x_i$, where $\varphi = (x_1, x_2, \ldots, x_n)$. Section 3.1.3 provides the definition of the derivative of f independent of the chart choice.

Similarly, we can define the differentiability of a map between manifolds M and N. A key fact here is that points on both manifolds can be represented by coordinates using charts. Note, however, that N is not necessarily covered by one chart, in contrast to the case of Definition 3.4 where $N = \mathbb{R}$.

Definition 3.5 Let M and N be m- and n-dimensional C^r manifolds, respectively. A map $f: M \to N$ is said to be of *class* C^0 if it is continuous. Let s be an integer satisfying $1 \leq s \leq r$. A continuous map $f: M \to N$ is said to be of *class* C^s at $p \in M$ if there exist C^r charts (U, φ) and (V, ψ) of M and N containing p and $f(p)$, respectively, such that $f(U) \subset V$ and $\psi \circ f \circ \varphi^{-1}: \varphi(U) \to \mathbb{R}^n$ is of class C^s at $\varphi(p)$. The map f is said to be of *class* C^s if it is of class C^s at any point on M.

In Definition 3.5, the composition $\psi \circ f \circ \varphi^{-1}$ is considered a local coordinate representation of f. Indeed, $\psi \circ f \circ \varphi^{-1}$ maps a local coordinate $\varphi(p) \in \mathbb{R}^m$ of $p \in M$ to $(\psi \circ f \circ \varphi^{-1})(\varphi(p)) = \psi(f(p)) \in \mathbb{R}^n$, which is a local coordinate of $f(p) \in N$.

Remark 3.1 In Definition 3.5, for charts (U, φ) of M and (V, ψ) of N and continuous map f, the preimage $f^{-1}(V)$ is open and $\left(U \cap f^{-1}(V), \varphi|_{U \cap f^{-1}(V)}\right)$ is also a chart of M, as we assume a maximal atlas. Therefore, even if $f(U) \not\subset V$, we can relabel $U \cap f^{-1}(V)$ as U to obtain $f(U) \subset V$.

Definition 3.6 Let I be an interval of $\mathbb{R}$ and M be a C^r manifold. Then, for s satisfying $0 \leq s \leq r$, a map $\gamma: I \to M$ of class C^s is called a C^s *curve*.

In Definition 3.6, we call the map itself, and not its image, a curve. However, depending on the context, we also use the term "curve" for the image of the map. The concept of a curve is used when defining tangent vectors in the next section.

Definition 3.7 Let M and N be C^r manifolds. A map $f: M \to N$ is called a C^s *diffeomorphism* if $f: M \to N$ is bijective and both f and f^{-1} are of class C^s. Then, we say that M and N are C^s *diffeomorphic* to each other.

In what follows, we suppose that $r = \infty$ and $s = \infty$, i.e., we consider C^∞ (or *smooth*) manifolds and focus on C^∞ (or *smooth*) functions and maps unless otherwise noted.

Remark 3.2 As a final remark in this subsection, we address the connectedness of manifolds. A topological space X is said to be *connected* if there exist no nonempty open subsets U and V such that $X = U \cup V$ and $U \cap V = \emptyset$. We say that X is *path-connected* if, for any $x, y \in X$, there exists a continuous curve $\gamma: [0, 1] \to X$ such that $\gamma(0) = x$ and $\gamma(1) = y$. If X is path-connected, then it is connected. A topological manifold M is connected if and only if it is path-connected. Throughout the book, if a manifold is not connected, then we focus on each of its connected component.

3.1.2 Tangent Vectors

A manifold is a generalized concept of surfaces. We define a vector space called a *tangent space* at each point on a smooth manifold as an analogy to a tangent plane at each point on a smooth surface.

We first review the Euclidean case. Consider a smooth curve $\gamma : I \to \mathbb{R}^n$ passing through a point $\gamma(0) =: p = (p_i) \in \mathbb{R}^n$, where I is an interval of $\mathbb{R}$ and contains 0. We can naturally express γ as

$$\gamma(t) = (x_1(t), x_2(t), \dots, x_n(t))^T = \sum_{i=1}^{n} x_i(t) e_i, \tag{3.1}$$

where all components of $e_i \in \mathbb{R}^n$ are 0 except for the ith component, which is 1, and $x_i(0) = p_i$. For this curve, because $\gamma(t) - \gamma(0) \in \mathbb{R}^n$, we can define

$$\gamma'(0) := \lim_{t \to 0} \frac{\gamma(t) - \gamma(0)}{t} = \frac{d}{dt} \gamma(t) \Big|_{t=0} = \sum_{i=1}^{n} \frac{dx_i}{dt}(0) e_i, \tag{3.2}$$

which is considered a tangent vector of the curve γ at $t = 0$.

However, the subtraction between $\gamma(0) \in M$ and $\gamma(t) \in M$ for a smooth curve γ on an n-dimensional manifold M is not defined in general. Although M can be related to $\mathbb{R}^n$ by charts, dependence on a specific chart does not work well. Indeed, for a chart (U, φ) around $\gamma(0) =: p \in M$, we can describe γ on a neighborhood of p as $\varphi(\gamma(t)) = (x_1(t), x_2(t), \dots, x_n(t))^T =: x(t) \in \mathbb{R}^n$ and compute $x'(0)$ as in (3.2); however, this depends on the chart choice.

To resolve this issue, we take an arbitrary smooth real-valued function f on M and consider the composition $f \circ \gamma : I \to \mathbb{R}$. Then, the derivative of $f \circ \gamma$ is conventional and independent of the chart choice. Nevertheless, we can write the derivative at $t = 0$ using a specific chart $(U, \varphi) = (U; x_1, x_2, \dots, x_n)$:

$$\frac{d}{dt} f(\gamma(t)) \Big|_{t=0} = \frac{d}{dt} (f \circ \varphi^{-1})(\varphi(\gamma(t))) \Big|_{t=0} = \sum_{i=1}^{n} \frac{dx_i}{dt}(0) \frac{\partial}{\partial x_i} \Big|_p f, \tag{3.3}$$

where $\frac{\partial}{\partial x_i}\big|_p f := \frac{\partial(f \circ \varphi^{-1})}{\partial x_i}(\varphi(p))$. We can regard $\frac{\partial}{\partial x_i}\big|_p$ as a map from $\mathfrak{F}_p(M)$ to $\mathbb{R}$, where $\mathfrak{F}_p(M)$ denotes the set of all smooth real-valued functions defined on a neighborhood of $p \in M$. Because the left-hand side of (3.3) is independent of φ, so too is $\sum_{i=1}^{n} \frac{dx_i}{dt}(0) \frac{\partial}{\partial x_i}\big|_p$. Since f is arbitrary, this quantity expresses the essence of a tangent vector on γ, which is a differential operator for real-valued functions on M.

Definition 3.8 Let M be a smooth manifold. A map $\xi : \mathfrak{F}_p(M) \to \mathbb{R}$ is called a *tangent vector* at a point p on M if there exists a curve γ on M such that $\gamma(0) = p$ and

$$\xi f = \frac{d}{dt} f(\gamma(t)) \Big|_{t=0}, \quad f \in \mathfrak{F}_p(M).$$

Then, we write ξ as $\dot{\gamma}(0)$. The set of all tangent vectors at p is shown to be an n-dimensional vector space, called the *tangent space* of M at p and denoted by $T_p M$.

In the aforementioned setting, we have

$$\dot{\gamma}(0) = \sum_{i=1}^{n} \frac{dx_i}{dt}(0) \left.\frac{\partial}{\partial x_i}\right|_p. \tag{3.4}$$

We again consider the case of $M = \mathbb{R}^n$ and recall γ in (3.1) and $\gamma'(0)$ in (3.2). It follows from Definition 3.8 and the chain rule that

$$\dot{\gamma}(0)f = \left.\frac{d}{dt}f(\gamma(t))\right|_{t=0} = Df(\gamma(0))[\gamma'(0)].$$

Note that the map $v \mapsto \xi$, where $v = \gamma'(0)$ and $\xi = \dot{\gamma}(0)$ for some curve γ, is an isomorphism (a linear bijection) between $\mathbb{R}^n$ and $T_{\gamma(0)}\mathbb{R}^n$, which is independent of the choice of γ. It is useful to identify $\gamma'(0) \in \mathbb{R}^n$ with $\dot{\gamma}(0) \in T_{\gamma(0)}\mathbb{R}^n$. Thus, we obtain the following relation, called the *canonical identification*:

$$T_p\mathbb{R}^n \simeq \mathbb{R}^n, \quad p \in \mathbb{R}^n.$$

In what follows, we use this identification when necessary. This identification, together with (3.2) and (3.4), implies that $e_i \in \mathbb{R}^n$ is identified with $\left.\frac{\partial}{\partial x_i}\right|_p \in T_p\mathbb{R}^n$ at any point $p \in \mathbb{R}^n$. Similarly, we can identify $T_X\mathbb{R}^{m \times n}$ with $\mathbb{R}^{m \times n}$ for any $X \in \mathbb{R}^{m \times n}$. This discussion can be further extended to a general vector space.

For a general smooth manifold M with a C^∞ atlas $\{(U_\lambda, \varphi_\lambda)\}_{\lambda \in \Lambda}$, let V be an arbitrary open set of M. We regard V as a topological space with the relative topology induced from the topology on M. Then, $\left\{\left(U_\lambda \cap V, \varphi_\lambda|_{U_\lambda \cap V}\right)\right\}_{\lambda \in \Lambda}$ is a C^∞ atlas of V, making V a smooth manifold called an *open submanifold* of M. Since a tangent vector at $p \in V \subset M$ to V is an operator acting on smooth functions defined on a neighborhood of p, we can identify $T_p V$ with $T_p M$ (see also Remark 3.4 in Sect. 3.1.4).

In addition, the set of all tangent vectors to an n-dimensional smooth manifold M, i.e., the disjoint union of all tangent spaces of M, defined as $TM := \bigcup_{p \in M} T_p M$, forms a manifold called the *tangent bundle* of M. The tangent bundle can be equipped with a C^∞ atlas and its dimension is $2n$. Specifically, for each chart (U, φ) of M, we let $TU := \bigcup_{p \in U} T_p U = \bigcup_{p \in U} T_p M$ and define $\tilde{\varphi}$ as follows: We can express $p \in U$ and $\xi \in T_p M$ using $\varphi = (x_1, x_2, \ldots, x_n)$ as $\varphi(p) \in \mathbb{R}^n$ and $\xi = \sum_{i=1}^{n} v_i \left.\frac{\partial}{\partial x_i}\right|_p$, with some $v = (v_i) \in \mathbb{R}^n$. Then, defining the map $\tilde{\varphi} \colon TU \to \varphi(U) \times \mathbb{R}^n$ by $\tilde{\varphi}(\xi) := (\varphi(p), v)$, we can show that $(TU, \tilde{\varphi})$ is a chart of TM. For a C^∞ atlas $\{(U_\lambda, \varphi_\lambda)\}_{\lambda \in \Lambda}$ of M, $\{(TU_\lambda, \tilde{\varphi}_\lambda)\}_{\lambda \in \Lambda}$ is then a C^∞ atlas of TM.

3.1.3 Derivatives of Maps

Thus far, we have discussed the differentiability of functions and maps. Next, we define their derivatives, which can be used in optimization algorithms on manifolds.

One of the important quantities in continuous optimization is the gradient of the objective function, if it exists. As defined in (2.10), each element of the Euclidean gradient of a function f on $\mathbb{R}^n$ is a partial derivative of f. For a function f defined on a manifold M, a similar quantity $\partial(f \circ \varphi^{-1})/\partial x_i$ with a chart $(U, \varphi) = (U; x_1, x_2, \ldots, x_n)$ around p clearly depends on the chart choice; this is undesirable.

In the Euclidean case, the gradient $\nabla f(p)$ of $f : \mathbb{R}^n \to \mathbb{R}$ at $p \in \mathbb{R}^n$ is closely related to the directional derivative $Df(p)[d]$ in the direction of $d \in \mathbb{R}^n$, as shown by (2.11), i.e., $Df(p)[d] = \langle \nabla f(p), d \rangle$, where $\langle \cdot, \cdot \rangle$ is the standard inner product. Thus, the operator $Df(p): \mathbb{R}^n \to \mathbb{R}$ acting as $d \mapsto Df(p)[d]$ is equal to $\langle \nabla f(p), \cdot \rangle$. Therefore, it may be possible to define the Riemannian gradient through the "derivative" of f, which is analogous to $Df(p)$ above, and an "inner product." We leave the definitions of the "inner product," which is defined via a Riemannian metric, and the Riemannian gradient for Sects. 3.2 and 3.3, respectively.

For any smooth curve γ on $\mathbb{R}^n$ with $\gamma(0) = p$, which is not necessarily a straight line, we have

$$Df(p)[\gamma'(0)] = \frac{d}{dt} f(\gamma(t)) \bigg|_{t=0} = (f \circ \gamma)'(0).$$

We exploit this relationship in the following discussion of the manifold case to define the derivatives of maps.

For a curve γ on a manifold M, a tangent vector $\dot{\gamma}(0)$ is defined. Let $f : M \to \mathbb{R}$ be a function on M. Note that $f \circ \gamma$ is a real-valued function of a real variable, so that $(f \circ \gamma)'(0)$ is still defined. We define the *derivative* $Df(p)$ of f at $p \in M$ as

$$Df(p)[\dot{\gamma}(0)] = (f \circ \gamma)'(0).$$

The derivative of a real-valued function is important for investigation of the objective function. We also require the derivative of a more general map between two manifolds for Riemannian optimization algorithms. One such example is a map called a retraction (see Sect. 3.4). From the above discussion, it is natural to define the derivative of a map as follows.

Definition 3.9 Let M and N be smooth manifolds. The *derivative* $DF(p): T_pM \to T_{F(p)}N$ of a smooth map $F: M \to N$ at $p \in M$ is defined, through curves γ on M with $\gamma(0) = p$ and $\gamma_F := F \circ \gamma$ on N, as

$$DF(p)[\dot{\gamma}(0)] = \dot{\gamma}_F(0). \tag{3.5}$$

It can be proved that the left-hand side of (3.5) is independent of the choice of γ (see (3.6)), and $DF(p)$ is well defined. The derivative $DF(p)$ maps a tangent

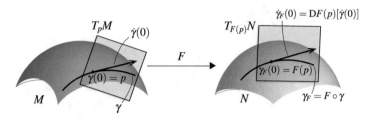

Fig. 3.2 Map $F: M \to N$ between manifolds M and N maps curve γ on M to curve $\gamma_F := F \circ \gamma$ on N. Its derivative $DF(p): T_pM \to T_{F(p)}N$ at $p \in M$ maps $\dot{\gamma}(0)$ to $DF(p)[\dot{\gamma}(0)] = \dot{\gamma}_F(0)$

vector $\dot{\gamma}(0) \in T_pM$ to $\dot{\gamma}_F(0)$, which is tangent to the curve $F \circ \gamma$ on N at $F(p)$ (see Fig. 3.2).

Remark 3.3 For the case of the Euclidean space $\mathbb{R}^n$ or a submanifold of the Euclidean space (see the next subsection), we use the same notation $DF(p)$ for both the directional derivative and the derivative defined in Definition 3.9. For example, for $F: \mathbb{R}^m \to \mathbb{R}^n$ and a curve γ with $\gamma(0) = p$ in $\mathbb{R}^m$, we identify $DF(p)[\dot{\gamma}(0)]$ with $DF(p)[\gamma'(0)]$, which are the derivative in Definition 3.9 and the directional derivative, respectively, by the canonical identification $T_{F(p)}\mathbb{R}^n \simeq \mathbb{R}^n$. Since we also identify $\dot{\gamma}(0)$ with $\gamma'(0)$ through the canonical identification $T_p\mathbb{R}^m \simeq \mathbb{R}^m$, use of D for both types of derivatives is justified. This useful notation is also used in Absil et al. (2008).

Let us see how the right-hand side of (3.5) acts on $g \in \mathfrak{F}_{F(p)}(N)$. We have

$$\dot{\gamma}_F(0)(g) = \frac{d(g \circ (F \circ \gamma))}{dt}\bigg|_{t=0} = \frac{d((g \circ F) \circ \gamma)}{dt}\bigg|_{t=0} = \dot{\gamma}(0)(g \circ F), \quad (3.6)$$

which implies that the right-hand side of (3.5) is independent of the choice of curve γ provided $\gamma(0) = p$ holds and $\dot{\gamma}(0)$ is the same tangent vector. Equations (3.5) and (3.6) give the following proposition.

Proposition 3.1 *Let M and N be smooth manifolds. The derivative $DF(p)$ of a smooth map $F: M \to N$ at $p \in M$ satisfies, for any $\xi \in T_pM$,*

$$(DF(p)[\xi])(g) = \xi(g \circ F), \quad g \in \mathfrak{F}_{F(p)}(N). \quad (3.7)$$

It is also possible to adopt (3.7) as the definition of $DF(p)$ instead of Definition 3.9.

The following formula for the derivative of the composition of two maps between manifolds is important and is used in many types of analyses.

Proposition 3.2 *Let M, N, and Q be smooth manifolds and $F: N \to Q$ and $G: M \to N$ be smooth maps. Then, for any $p \in M$ and $\xi \in T_pM$, we have*

$$D(F \circ G)(p)[\xi] = DF(G(p))[DG(p)[\xi]].$$

Proof It follows from Proposition 3.1 that, for any $h \in \mathfrak{F}_{F(G(p))}(Q)$,

$$(D(F \circ G)(p)[\xi])(h) = \xi(h \circ F \circ G) = (DG(p)[\xi])(h \circ F)$$
$$= (DF(G(p))[DG(p)[\xi]])(h).$$

Because h is arbitrary, the claim is proved. $\square$

3.1.4 Embedded Submanifolds

The sphere S^{n-1} is one of the most basic manifolds we consider in this book. Note that S^{n-1} is clearly a subset of $\mathbb{R}^n$. We clarify the relationship between S^{n-1} and $\mathbb{R}^n$ to gain an insight into whether S^{n-1} can be equipped with a manifold structure inherited from $\mathbb{R}^n$. To this end, we introduce the notion of a *submanifold*.

Definition 3.10 Let M be an m-dimensional manifold and N be a subset of M. The subset $N \subset M$ is called an n-dimensional *(regular) submanifold* of M if, for any $p \in N$, there exists a chart $(U; x_1, x_2, \ldots, x_m)$ containing p such that the intersection $N \cap U$ is written as $N \cap U = \{q \in U \mid x_{n+1}(q) = x_{n+2}(q) = \cdots = x_m(q) = 0\}$.

It can be shown that a regular submanifold N of M is indeed a manifold with an atlas consisting of charts, each of which is induced from the corresponding chart of M described in the above definition. If $m = n$, then the regular submanifold N is an open submanifold of M.

Example 3.3 The sphere $S^2 = \{x \in \mathbb{R}^3 \mid x^T x = 1\}$ is a regular submanifold of $\mathbb{R}^3$. To prove this statement, we let $p = (p_1, p_2, p_3)^T \in S^2$ be an arbitrary point and construct a chart containing p that satisfies the condition in Definition 3.10. At least one among $p_1, p_2,$ and p_3 is nonzero; thus, it is sufficient to discuss the case of $p_3 > 0$. Let $U := \{x = (x_1, x_2, x_3)^T \in \mathbb{R}^3 \mid x_1^2 + x_2^2 < 1, x_3 > 0\}$ and define $\varphi: U \to \mathbb{R}^3$ by $\varphi(x) := \left(x_1, x_2, x_3 - \sqrt{1 - x_1^2 - x_2^2}\right)^T =: (y_1, y_2, y_3)^T$. Then, $(U, \varphi) = (U; y_1, y_2, y_3)$ is a chart of $\mathbb{R}^3$ around p and we have $S^2 \cap U = \{q \in U \mid y_3(q) = 0\}$. The other cases can be similarly discussed. Therefore, S^2 is a regular submanifold of $\mathbb{R}^3$.

Similarly, S^{n-1} is shown to be a regular submanifold of $\mathbb{R}^n$.

Although the discussion in this example is not complicated, determining whether a subset of a manifold is a regular submanifold may be more difficult in more general cases. It will be convenient if we can prove that some specific subset of a manifold is a submanifold without using local coordinates. A map called an *embedding* plays an important role in characterizing a regular submanifold.

Definition 3.11 Let M and N be smooth manifolds. A smooth map $f: M \to N$ is called an *immersion* if $Df(p): T_pM \to T_{f(p)}N$ is injective for any $p \in M$. An immersion $f: M \to N$ is called an *embedding* if $f: M \to f(M)$ is a homeomorphism, where $f(M) \subset N$ is endowed with the relative topology.

The following theorem can be proved using the inverse function theorem. We omit proofs of both theorems. Interested readers can refer to, e.g., Tu (2011).

Theorem 3.1 *Let M and N be smooth manifolds. If N is a regular submanifold of M, then the inclusion map $\iota\colon N \to M\colon p \mapsto p$ is an embedding. Conversely, if there exists an embedding $f\colon M \to N$, then the image $f(M)$ is an n-dimensional regular submanifold of M.*

In view of this theorem, a regular submanifold is also called an *embedded submanifold*; hereafter, we refer to it simply as a *submanifold*.

Definition 3.12 Let M and N be smooth manifolds and $f\colon M \to N$ be a smooth map between them. A point $p \in M$ is called a *regular point* of f if the derivative $Df(p)\colon T_pM \to T_{f(p)}N$ is surjective. If p is not a regular point, p is called a *critical point*. A point $q \in N$ is called a *critical value* of f if there exists a critical point $p \in M$ of f such that $f(p) = q$. If q is not a critical value, q is called a *regular value*.

By definition, a point $q \in N$ is a regular value of $f\colon M \to N$ if and only if any $p \in f^{-1}(\{q\})$ is a regular point of f.

The following theorem is a powerful tool for proving that a subset of a particular type is a submanifold. See also Tu (2011) for its proof.

Theorem 3.2 *Assume $m \geq n > 0$ and let M and N be m- and n-dimensional smooth manifolds, respectively. For a smooth map $f\colon M \to N$, if $q \in N$ is a regular value of f and $f^{-1}(\{q\}) \neq \emptyset$, then $f^{-1}(\{q\})$ is an $(m - n)$-dimensional submanifold of M.*

Assume $p \leq n$. The set $\mathrm{St}(p, n) := \{X \in \mathbb{R}^{n \times p} \mid X^T X = I_p\}$ is very important because numerous optimization problems have the constraint $X^T X = I_p$ on their decision variable $X \in \mathbb{R}^{n \times p}$, i.e., column vectors of X are orthonormal with respect to the standard inner product in $\mathbb{R}^n$. We wish to show that $\mathrm{St}(p, n)$ can be equipped with a manifold structure to apply optimization algorithms to problems on $\mathrm{St}(p, n)$ that are analogous to that discussed in Sect. 1.1.2. For clearer understanding, we discuss $\mathrm{St}(2, n)$ before proceeding to a general discussion of $\mathrm{St}(p, n)$.

Example 3.4 Consider the set $S(2, n) := \{(x, y) \in \mathbb{R}^n \times \mathbb{R}^n \mid x^T x = y^T y = 1, x^T y = 0\}$ with $n \geq 2$. Noting that the matrix $X = (x, y) \in \mathbb{R}^{n \times 2}$ belongs to $\mathrm{St}(2, n)$ if and only if the columns x and y satisfy $(x, y) \in S(2, n)$, we identify $S(2, n)$ with $\mathrm{St}(2, n)$.

We define $f\colon \mathbb{R}^n \times \mathbb{R}^n \to \mathbb{R}^3$ as $f(x, y) := (x^T x, y^T y, x^T y)^T$ to obtain the expression $\mathrm{St}(2, n) = f^{-1}(\{(1, 1, 0)^T\})$. To show that $\mathrm{St}(2, n)$ is a $(2n - 3)$-dimensional submanifold of $\mathbb{R}^n \times \mathbb{R}^n$, it is sufficient to prove that $(1, 1, 0)^T$ is a regular value of f, because of Theorem 3.2. We identify $\mathbb{R}^n \times \mathbb{R}^n$ with $\mathbb{R}^{2n}$ and express $f(x, y)$ as $(w_1, w_2, w_3)^T := f(x, y) = f(x_1, x_2, \ldots, x_n, y_1, y_2, \ldots, y_n) =: f(z_1, z_2, \ldots, z_{2n})$. The derivative $Df(x, y)$ can be expressed by the *Jacobian matrix* $(Jf)_{(x,y)} \in \mathbb{R}^{3 \times 2n}$, whose (i, j) element is defined as $\partial w_i / \partial z_j$. We then have $(Jf)_{(x,y)} = \begin{pmatrix} 2x & 0 & y \\ 0 & 2y & x \end{pmatrix}^T$. If $a_1(2x^T, 0) + a_2(0, 2y^T) + a_3(y^T, x^T) = 0$ for

$a_1, a_2, a_3 \in \mathbb{R}$, then it holds that $2a_1 x + a_3 y = 2a_2 y + a_3 x = 0$. It follows from $x^T x = y^T y = 1$ and $x^T y = y^T x = 0$ for $(x, y) \in \text{St}(2, n)$ that $a_1 = a_2 = a_3 = 0$, implying that the three row vectors of $(Jf)_{(x,y)}$ are linearly independent. Therefore, $(Jf)_{(x,y)}$, and hence $Df(x, y)$, are surjective for any $(x, y) \in \text{St}(2, n) = f^{-1}(\{(1, 1, 0)^T\})$. It follows that all points of $f^{-1}(\{(1, 1, 0)^T\})$ are regular points, and therefore $(1, 1, 0)^T$ is a regular value of f.

Example 3.5 We can show that $\text{St}(p, n)$ is an $(np - p(p + 1)/2)$-dimensional submanifold of $\mathbb{R}^{n \times p}$ in a similar manner to the previous example. In contrast, we can also prove this statement without explicitly writing the Jacobian matrix. The following discussion is based on that in Absil et al. (2008).

The condition $X^T X = I_p$ for $X = (x_1, x_2, \ldots, x_p) \in \mathbb{R}^{n \times p}$ consists of $p(p + 1)/2$ equalities $x_i^T x_i = 1$ for $i = 1, 2, \ldots, p$ and $x_i^T x_j = 0$ for $i, j = 1, 2, \ldots, p$ with $i < j$, where the conditions $x_i^T x_j = 0$ for $i, j = 1, 2, \ldots, p$ with $i > j$ are duplicated and should be ignored when counting the number of constraints. This duplication stems from the fact that $X^T X$ is a symmetric matrix. We define the map $F \colon \mathbb{R}^{n \times p} \to \text{Sym}(p)$ by $F(X) := X^T X$ to obtain $\text{St}(p, n) = F^{-1}(\{I_p\})$. We need to prove that $DF(X)$ is surjective at any $X \in F^{-1}(\{I_p\})$ and, thus, I_p is a regular value of F; this, together with Theorem 3.2, yields the desired claim. The derivative $DF(X) \colon T_X \mathbb{R}^{n \times p} \to T_{F(X)} \text{Sym}(p)$, regarded as $DF(X) \colon \mathbb{R}^{n \times p} \to \text{Sym}(p)$ (see Remark 3.4), acts on $Y \in \mathbb{R}^{n \times p}$ as $DF(X)[Y] = X^T Y + Y^T X$. Let X be an arbitrary element of $F^{-1}(\{I_p\})$, i.e., $X^T X = I_p$. Then, for any $Z \in \text{Sym}(p)$, $Y = XZ/2$ satisfies $DF(X)[Y] = (X^T X Z + Z X^T X)/2 = Z$. Hence, $DF(X)$ is surjective.

Definition 3.13 The set $\text{St}(p, n) = \{X \in \mathbb{R}^{n \times p} \mid X^T X = I_p\}$, as a submanifold of $\mathbb{R}^{n \times p}$ as discussed in Example 3.5, is called the *(compact) Stiefel manifold*.

As stated previously, the Stiefel manifold is very important because it has a wide variety of applications (see Chap. 5).

Remark 3.4 Let N be a submanifold of a manifold M. The tangent space $T_x N$ at a point x on N is expressed as a linear subspace of $T_x M$. Indeed, the inclusion map $\iota \colon N \to M$ satisfies $D\iota(x)[\xi] \in T_x M$ for $\xi \in T_x N$. For any smooth function $\bar{f} \colon M \to \mathbb{R}$ and its restriction $f := \bar{f}|_N \colon N \to \mathbb{R}$, we have

$$(D\iota(x)[\xi])\bar{f} = \xi(\bar{f} \circ \iota) = \xi f.$$

Thus, we identify ξ with $D\iota(x)[\xi]$ and, thereby, $T_x N$ with a subspace of $T_x M$. Specifically, we can identify $T_x M$ with a subspace of $T_x \mathbb{R}^n \simeq \mathbb{R}^n$ if M is a submanifold of $\mathbb{R}^n$, and $T_X M$ with a subspace of $T_X \mathbb{R}^{m \times n} \simeq \mathbb{R}^{m \times n}$ if M is a submanifold of $\mathbb{R}^{m \times n}$.

Using $T_X \text{St}(p, n) \subset T_X \mathbb{R}^{n \times p} \simeq \mathbb{R}^{n \times p}$, for the map F in Example 3.5, we have $DF(X)[\xi] = \frac{d}{dt} F(\gamma(t))|_{t=0} = 0$ for any $\xi \in T_X \text{St}(p, n)$ and curve γ on $\text{St}(p, n)$ with $\gamma(0) = X$ and $\dot{\gamma}(0) = \xi$, because $F(\gamma(t)) = I_p$ for all t. Therefore, the relation $T_X \text{St}(p, n) \subset DF(X)^{-1}(\{0\})$ holds. The dimension of the kernel $DF(X)^{-1}(\{0\})$ of $DF(X)$, which is surjective, can be counted using the rank–nullity theorem in linear algebra as $\dim \mathbb{R}^{n \times p} - \dim \text{Sym}(p) = np - p(p + 1)/2 = \dim \text{St}(p, n)$. Thus,

$T_X \operatorname{St}(p, n) \, (\subset DF(X)^{-1}(\{0\}))$ and $DF(X)^{-1}(\{0\})$ coincide because they are vector spaces of the same dimension. To summarize, the tangent space $T_X \operatorname{St}(p, n)$ is given by

$$T_X \operatorname{St}(p, n) = DF(X)^{-1}(\{0\}) = \{\xi \in \mathbb{R}^{n \times p} \mid X^T \xi + \xi^T X = 0\}. \tag{3.8}$$

In particular, for the sphere $S^{n-1} = \operatorname{St}(1, n)$, we have

$$T_x S^{n-1} = \{\xi \in \mathbb{R}^n \mid x^T \xi = 0\}. \tag{3.9}$$

Another expression of $T_X \operatorname{St}(p, n)$ is possible when we use a matrix $X_\perp \in \mathbb{R}^{n \times (n-p)}$ that satisfies $X_\perp^T X_\perp = I_{n-p}$ and $X^T X_\perp = 0$. In other words, the $n - p$ columns of $X_\perp$ form an orthonormal basis of the orthogonal complement of the p-dimensional subspace spanned by the p columns of X in $\mathbb{R}^n$ with respect to the standard inner product. Such $X_\perp$ is not determined to be unique, and any $X_\perp$ can be used. Since the columns of X and $X_\perp$ span $\mathbb{R}^n$, any vector in $\mathbb{R}^n$ can be expressed by a linear combination of the columns of X and $X_\perp$. Similarly, any $n \times p$ matrix can be expressed by using X and $X_\perp$, as each column of the $n \times p$ matrix is an n-dimensional vector. Consequently, for $\xi \in T_X \operatorname{St}(p, n) \subset \mathbb{R}^{n \times p}$, there exist matrices $B \in \mathbb{R}^{p \times p}$ and $C \in \mathbb{R}^{(n-p) \times p}$ such that $\xi = XB + X_\perp C$. In this case, the condition $X^T \xi + \xi^T X = 0$ shown in (3.8) is equivalent to $B + B^T = 0$, i.e., B is a skew-symmetric matrix. Therefore, we obtain the expression

$$T_X \operatorname{St}(p, n) = \left\{ XB + X_\perp C \in \mathbb{R}^{n \times p} \mid B \in \operatorname{Skew}(p), \ C \in \mathbb{R}^{(n-p) \times p} \right\}. \tag{3.10}$$

3.1.5 Quotient Manifolds

We do not discuss quotient manifolds in detail in this book. Here, we simply introduce the concept and provide an example of quotient manifolds.

The set of all p-dimensional subspaces of $\mathbb{R}^n$ for fixed $p \leq n$ is denoted by $\operatorname{Grass}(p, n)$.[1] It is shown that $\operatorname{Grass}(p, n)$ can be a $p(n - p)$-dimensional manifold, called the *Grassmann manifold*. A point on $\operatorname{Grass}(p, n)$ is a subspace of $\mathbb{R}^n$. When handling such objects, the concept of a quotient manifold plays an important role.

The columns of each point X on the Stiefel manifold $\operatorname{St}(p, n)$ form an orthonormal basis of a p-dimensional subspace of $\mathbb{R}^n$, i.e., a point on $\operatorname{Grass}(p, n)$. When $X_1, X_2 \in \operatorname{St}(p, n)$ are related with $X_2 = X_1 Q$ for some $Q \in \mathcal{O}_p$, the columns of X_1 and X_2 are orthonormal bases of the same subspace. Hence, we define an equivalence relation $\sim$ on $\operatorname{St}(p, n)$ as $X_1 \sim X_2$ if and only if there exists $Q \in \mathcal{O}_p$ such that $X_2 = X_1 Q$, to identify X_1 with X_2. Then, a subspace spanned by the columns of $X \in \operatorname{St}(p, n)$ is identified with the equivalence class $[X]$ and $\operatorname{Grass}(p, n)$ can be identified with $\operatorname{St}(p, n)/\sim$. Further, $\mathcal{O}_p$ is a group and $\phi \colon \operatorname{St}(p, n) \times \mathcal{O}_p \to \operatorname{St}(p, n)$

---

[1] As another approach, we can identify $\operatorname{Grass}(p, n)$ with $\{X \in \mathbb{R}^{n \times n} \mid X^2 = X = X^T, \operatorname{tr}(X) = p\}$, which is proved to be a submanifold of $\mathbb{R}^{n \times n}$ (Bodmann and Haas 2015; Sato and Iwai 2014).

as $(X, Q) \mapsto XQ$ is a right action. Thus, we have $\text{Grass}(p, n) = \text{St}(p, n)/\mathcal{O}_p$ (see also Sect. 2.4).

For a general manifold M with an equivalence relation $\sim$, suppose that $M/\sim$ is also a manifold (with an appropriate atlas). The manifold $M/\sim$ is called a *quotient manifold* of M if the natural projection $\pi: M \to M/\sim$ is a *submersion*, i.e., the derivative $D\pi(x)$ is surjective at each $x \in M$. It is known that $\text{Grass}(p, n) = \text{St}(p, n)/\mathcal{O}_p$ can be a quotient manifold of $\text{St}(p, n)$. For further details, see Absil et al. (2008), Lee (2012).

An advantage of Riemannian optimization theory is that it can handle optimization on general Riemannian manifolds such as $\text{Grass}(p, n)$, which are not necessarily expressed as submanifolds of Euclidean spaces.

3.2 Riemannian Metric and Riemannian Manifolds

To discuss inner products of tangent vectors, which are important in optimization, we introduce the concepts of Riemannian metric and Riemannian manifold.

Definition 3.14 Let M be a smooth manifold. For any $x \in M$, let $g_x: T_x M \times T_x M \to \mathbb{R}$ be an inner product on $T_x M$. If $g: x \mapsto g_x$ is smooth, g is called a *Riemannian metric* on M and (M, g) is called a *Riemannian manifold*. If g is clear from the context, we simply say that M is a Riemannian manifold.

Since g_x for each $x \in M$ is an inner product on $T_x M$, we also use $\langle \xi, \eta \rangle_x$ to denote $g_x(\xi, \eta)$ for $\xi, \eta \in T_x M$. Furthermore, since $T_x M$ is, thus, an inner product space, we can define the norm of $\xi \in T_x M$ as $\|\xi\|_x := \sqrt{\langle \xi, \xi \rangle_x}$ (see (2.4) in Example 2.3).

Theorem 3.3 *For any smooth manifold M, there exists a Riemannian metric on M.*

The second-countability of M is fundamental for proving Theorem 3.3. Although we omit a detailed proof of this theorem, it can be briefly described as follows: Any manifold M with second-countability, as in Definition 3.1, is shown to be *paracompact*, which is a weaker condition than *compactness*. It then follows that M has a *partition of unity* subordinate to any open covering, i.e., any family of open sets $\{U_\lambda\}_{\lambda \in \Lambda}$ in M such that $\cup_{\lambda \in \Lambda} U_\lambda = M$. Theorem 3.3 can be proved using a partition of unity subordinate to an open covering $\{U_\alpha\}$, where $(U_\alpha, \varphi_\alpha)$ are smooth charts belonging to the atlas of M. For more detail, see, e.g., Lee (2012).

Example 3.6 Let N be a submanifold of a Riemannian manifold (M, g^M). We can define a Riemannian metric g^N on N as, for any $x \in N \subset M$ and $\xi, \eta \in T_x N \subset T_x M$,

$$g_x^N(\xi, \eta) := g_x^M(\xi, \eta).$$

Strictly speaking, the right-hand side is $g_{\iota(x)}^M(D\iota(x)[\xi], D\iota(x)[\eta])$ with the inclusion map $\iota: N \to M$. The Riemannian metric g^N and manifold (N, g^N) thus defined are called an *induced metric* on N and *Riemannian submanifold* of M, respectively.

Example 3.7 The Euclidean space $\mathbb{R}^n$ with the standard inner product (2.2) can be naturally regarded as a Riemannian manifold; the Riemannian metric in this case is defined by, for every point $x \in \mathbb{R}^n$, $\langle \xi, \eta \rangle_x := \xi^T \eta$ for $\xi, \eta \in T_x \mathbb{R}^n \simeq \mathbb{R}^n$. In other words, we regard ξ and η as vectors in $\mathbb{R}^n$ and define their inner product as that by (2.2). The sphere $S^{n-1} = \{x \in \mathbb{R}^n \mid x^T x = 1\}$ is a Riemannian submanifold of the Riemannian manifold $\mathbb{R}^n$ with the induced metric

$$\langle \xi, \eta \rangle_x := \xi^T \eta, \quad \xi, \eta \in T_x S^{n-1} \subset T_x \mathbb{R}^n \simeq \mathbb{R}^n, \ x \in S^{n-1}. \tag{3.11}$$

Similarly, the matrix space $\mathbb{R}^{n \times p}$ can be regarded as a Riemannian manifold with the Riemannian metric such that, for every $X \in \mathbb{R}^{n \times p}$, $\langle \xi, \eta \rangle_X := \mathrm{tr}(\xi^T \eta)$ for arbitrary $\xi, \eta \in T_X \mathbb{R}^{n \times p} \simeq \mathbb{R}^{n \times p}$. In particular, when $p \leq n$, the Stiefel manifold $\mathrm{St}(p, n) = \{X \in \mathbb{R}^{n \times p} \mid X^T X = I_p\}$ is a Riemannian submanifold of $\mathbb{R}^{n \times p}$ with the induced metric

$$\langle \xi, \eta \rangle_X := \mathrm{tr}(\xi^T \eta), \quad \xi, \eta \in T_X \mathrm{St}(p, n) \subset T_X \mathbb{R}^{n \times p} \simeq \mathbb{R}^{n \times p}, \ X \in \mathrm{St}(p, n). \tag{3.12}$$

A Riemannian metric enables us to define various geometric concepts such as angles and distances on manifolds. Furthermore, Riemannian manifolds are the focus of the discussion of optimization presented herein; i.e., we are interested in the following *unconstrained optimization problem* with an objective function $f : M \to \mathbb{R}$ on a general Riemannian manifold M.

Problem 3.1

$$\begin{aligned} \text{minimize} \quad & f(x) \\ \text{subject to} \quad & x \in M. \end{aligned}$$

Here, *global* and *local optimal solutions* to Problem 3.1 are defined in the same manner as the Euclidean case by replacing $\mathbb{R}^n$ in Definition 2.18 with M.

3.3 Gradients of Functions on Riemannian Manifolds

As discussed in Sect. 2.6, it seems natural to define the gradient of a function $f : \mathbb{R}^n \to \mathbb{R}$ depending on the choice of inner product in $\mathbb{R}^n$. Thus, we define the gradient of a function on a Riemannian manifold as follows.

Definition 3.15 For a smooth real-valued function f defined on a smooth Riemannian manifold (M, g), the *(Riemannian) gradient* of f at $x \in M$ is defined as a unique tangent vector $\mathrm{grad}\, f(x)$ on M at x that satisfies

$$\mathrm{D}f(x)[\xi] = g_x(\mathrm{grad}\, f(x), \xi), \quad \xi \in T_x M. \tag{3.13}$$

The statement that grad $f(x)$ in Definition 3.15 is unique can be easily checked, as follows. Assume that $x \in M$ and that $\eta, \zeta \in T_x M$ satisfy the condition to be the gradient of f at x: $Df(x)[\xi] = g_x(\eta, \xi) = g_x(\zeta, \xi)$ for any $\xi \in T_x M$. Then, we have $g_x(\eta - \zeta, \xi) = 0$. Note that $\eta - \zeta$ is also in $T_x M$ because $T_x M$ is a vector space. As ξ is arbitrary, the above equation means that $\eta - \zeta = 0$, i.e., $\eta = \zeta$.

The discussion in Sect. 2.6 can be justified in the context of the Riemannian setting. For example, we observed that $G^{-1} \nabla f(x)$ can be an appropriate gradient of f in $\mathbb{R}^n$ equipped with the inner product (2.1), where $G \in \mathrm{Sym}(n)$ is positive definite. We now review this observation from the Riemannian perspective. The inner product space $\mathbb{R}^n$ with (2.1) is identified with a Riemannian manifold $\mathbb{R}^n$ endowed with the following Riemannian metric g:

$$g_x(a, b) := a^T G b, \quad a, b \in T_x \mathbb{R}^n \simeq \mathbb{R}^n, \ x \in \mathbb{R}^n.$$

The vector $G^{-1} \nabla f(x)$ satisfies $g_x(G^{-1} \nabla f(x), \xi) = \nabla f(x)^T \xi = Df(x)[\xi]$ for any $\xi \in \mathbb{R}^n \simeq T_x \mathbb{R}^n$, where $\nabla f(x)$ is the Euclidean gradient. By definition, this means that the Riemannian gradient is expressed as grad $f(x) = G^{-1} \nabla f(x)$.

Example 3.8 For a given vector $c \in \mathbb{R}^n$, consider the function $f(x) := c^T x$ defined on S^{n-1}, which we regard as a Riemannian manifold with Riemannian metric (3.11).

First, let us compute grad $f(x)$ at $x \in S^{n-1}$ using a chart. Without loss of generality, we assume $x \in U_n^+ := \{x = (x_i) \in S^{n-1} \mid x_n > 0\}$. As in Example 3.2, we consider a chart (U_n^+, φ_n^+) with $\varphi_n^+(x) := \bar{x} := (x_1, x_2, \ldots, x_{n-1})^T \in \mathbb{R}^{n-1}$ for $x \in U_n^+ \subset \mathbb{R}^n$. Denoting $y = (y_1, y_2, \ldots, y_{n-1})^T := \varphi_n^+(x)$, we can write grad $f(x) \in T_x S^{n-1}$ as grad $f(x) = \sum_{i=1}^{n-1} g_i \frac{\partial}{\partial y_i}$, where $g = (g_i) \in \mathbb{R}^{n-1}$ and $\frac{\partial}{\partial y_i}\big|_x$ is denoted by $\frac{\partial}{\partial y_i}$ for simplicity. Noting $x_i = y_i$ for $i = 1, 2, \ldots, n-1$ and $x_n = \sqrt{1 - \bar{x}^T \bar{x}} = \sqrt{1 - y^T y}$, and identifying $\frac{\partial}{\partial y_i} \in T_x S^{n-1}$ with $D\iota(x)\left[\frac{\partial}{\partial y_i}\right] \in T_x \mathbb{R}^n$, we have

$$\frac{\partial}{\partial y_i} = \sum_{j=1}^{n} \frac{\partial x_j}{\partial y_i} \frac{\partial}{\partial x_j} = \frac{\partial}{\partial x_i} - \frac{x_i}{x_n} \frac{\partial}{\partial x_n}. \tag{3.14}$$

The induced metric (3.11) yields $\left\langle \frac{\partial}{\partial x_i}, \frac{\partial}{\partial x_j} \right\rangle_x = \delta_{ij}$ for $i, j = 1, 2, \ldots, n$, which, together with (3.14), implies $\left\langle \frac{\partial}{\partial y_i}, \frac{\partial}{\partial y_j} \right\rangle_x = \delta_{ij} + \frac{x_i x_j}{x_n^2}$ for $i, j = 1, 2, \ldots, n-1$. Defining $G := \left(\left\langle \frac{\partial}{\partial y_i}, \frac{\partial}{\partial y_j} \right\rangle_x \right) = I_{n-1} + \frac{\bar{x} \bar{x}^T}{x_n^2} \in \mathrm{Sym}(n-1)$ and $\bar{c} := (c_1, c_2, \ldots, c_{n-1})^T \in \mathbb{R}^{n-1}$, for any $\xi = \sum_{i=1}^{n-1} v_i \frac{\partial}{\partial y_i} \in T_x S^{n-1}$ with $v = (v_i) \in \mathbb{R}^{n-1}$, we have

$$\langle \mathrm{grad}\, f(x), \xi \rangle_x = \left\langle \sum_{i=1}^{n-1} g_i \frac{\partial}{\partial y_i}, \sum_{j=1}^{n-1} v_j \frac{\partial}{\partial y_j} \right\rangle_x = \sum_{i=1}^{n-1} \sum_{j=1}^{n-1} g_i v_j \left\langle \frac{\partial}{\partial y_i}, \frac{\partial}{\partial y_j} \right\rangle_x = v^T G g$$

and

$$\mathrm{D}f(x)[\xi] = \xi f = \sum_{i=1}^{n-1} v_i \frac{\partial}{\partial y_i} \left(\sum_{j=1}^{n-1} c_j y_j + c_n \sqrt{1 - y^T y} \right) = v^T \left(\bar{c} - \frac{c_n}{\sqrt{1 - y^T y}} y \right),$$

which are equal to each other based on the definition of the Riemannian gradient (3.13). Because ξ and, hence, v are arbitrary, we have $Gg = \bar{c} - c_n y / \sqrt{1 - y^T y}$. A simple calculation gives $G^{-1} = I_{n-1} - yy^T$, implying

$$g = G^{-1} \left(\bar{c} - \frac{c_n}{\sqrt{1 - y^T y}} y \right) = \bar{c} - (\bar{c}^T y + c_n \sqrt{1 - y^T y}) y. \tag{3.15}$$

Thus, we obtain the following expression for grad $f(x)$:

$$\mathrm{grad}\, f(x) = \sum_{i=1}^{n-1} g_i \frac{\partial}{\partial y_i} = \sum_{i=1}^{n-1} \left(c_i - \left(\sum_{j=1}^{n-1} c_j y_j + c_n \left(1 - \sum_{j=1}^{n-1} y_j^2 \right)^{1/2} \right) y_i \right) \frac{\partial}{\partial y_i}. \tag{3.16}$$

This expression contains quantities depending on the chart choice and may be undesirable. In fact, we can further rewrite (3.16) using the coordinate in $\mathbb{R}^n$. To this end, we identify $\frac{\partial}{\partial x_i} \in T_x \mathbb{R}^n$ with $e_i \in \mathbb{R}^n$ to obtain

$$\mathrm{grad}\, f(x) = \sum_{i=1}^{n-1} g_i \frac{\partial}{\partial y_i} = \sum_{i=1}^{n-1} g_i \left(\frac{\partial}{\partial x_i} - \frac{x_i}{x_n} \frac{\partial}{\partial x_n} \right)$$

$$= \left(g_1, g_2, \ldots, g_{n-1}, -\frac{g^T \bar{x}}{x_n} \right)^T \in \mathbb{R}^n.$$

Because $y = \bar{x}$ and $\sqrt{1 - y^T y} = x_n$, from (3.15), we have $g = \bar{c} - (c^T x)\bar{x}$. It follows that $g^T \bar{x} = -c_n x_n + (c^T x) x_n^2$ and

$$\mathrm{grad}\, f(x) = ((\bar{c} - (c^T x)\bar{x})^T, c_n - (c^T x) x_n)^T = c - (c^T x)x. \tag{3.17}$$

This expression is based on the coordinate in $\mathbb{R}^n$ and is independent of the chart choice. Thus, it may be more suitable for optimization algorithms than (3.16).

    The above computation may seem tedious even though the objective function f has a simple form. Here, we take another, much simpler approach to finding the expression of grad $f(x)$. We define $\bar{f} \colon \mathbb{R}^n \to \mathbb{R}$ as $\bar{f}(x) := c^T x$, the restriction of which to S^{n-1} is equal to f. The Euclidean gradient $\nabla \bar{f}(x) = c$ satisfies $\mathrm{D}\bar{f}(x)[\xi] = \nabla \bar{f}(x)^T \xi$ for all $\xi \in T_x S^{n-1}$. Since $\nabla \bar{f}(x)$ is not in $T_x S^{n-1}$ in general, we cannot conclude that this is the Riemannian gradient. However, we consider the Riemannian metric induced from the standard inner product in $\mathbb{R}^n$; thus, it seems that we can obtain the Riemannian gradient grad $f(x)$ by modifying the Euclidean gradient $\nabla \bar{f}(x)$. A key aspect is the decomposition of $\mathbb{R}^n$ into $T_x S^{n-1}$ and $(T_x S^{n-1})^\perp$, where $(T_x S^{n-1})^\perp$ is the orthogonal complement of $T_x S^{n-1}$ in $\mathbb{R}^n$ with respect to the standard inner product. Accordingly, we can decompose $\nabla \bar{f}(x) \in \mathbb{R}^n$ into two components: $\zeta_x \in$

$T_x S^{n-1} = \{\xi \in \mathbb{R}^n \mid x^T \xi = 0\}$ (see (3.9)) and $d_x \in (T_x S^{n-1})^{\perp} = \{\alpha x \mid \alpha \in \mathbb{R}\}$, as

$$\nabla \bar{f}(x) = c = \zeta_x + d_x = (I_n - xx^T)c + x(x^T c). \tag{3.18}$$

Then, we can observe that $d_x = x(x^T c) \in (T_x S^{n-1})^{\perp}$ in (3.18) clearly does not affect the value of $\nabla \bar{f}(x)^T \xi$, i.e., $d_x^T \xi = 0$ for any $\xi \in T_x S^{n-1}$. Therefore, we have $D\bar{f}(x)[\xi] = \zeta_x^T \xi$ for $\zeta_x \in T_x S^{n-1}$, which, of course, yields the same result as (3.17):

$$\operatorname{grad} f(x) = \zeta_x = (I_n - xx^T)c = c - (c^T x)x.$$

The above example implies that we can easily obtain the Riemannian gradient on a Riemannian manifold M when M is a submanifold of a Euclidean space. More generally, for a submanifold M of another ambient Riemannian manifold $(\bar{M}, \bar{g})$, we have the following proposition. Here, in the tangent space $T_x \bar{M}$ at $x \in \bar{M}$, the orthogonal complement of $T_x M \subset T_x \bar{M}$ with respect to the inner product defined by $\bar{g}_x$ is called the *normal space* and is denoted by $N_x M$, i.e.,

$$N_x M := \{\eta \in T_x \bar{M} \mid \bar{g}_x(\eta, \xi) = 0 \text{ for all } \xi \in T_x M\}.$$

Proposition 3.3 *Let $(\bar{M}, \bar{g})$ be a smooth Riemannian manifold and let (M, g) be its Riemannian submanifold. Let $P_x : T_x \bar{M} \to T_x M$ be the orthogonal projection onto $T_x M$; i.e., for any $\chi \in T_x \bar{M}$, it holds that $P_x(\chi) \in T_x M \subset T_x \bar{M}$ and $\chi - P_x(\chi) \in N_x M \subset T_x \bar{M}$. Further, assume that the restriction of a smooth function $\bar{f} : \bar{M} \to \mathbb{R}$ to M is $f : M \to \mathbb{R}$. Then, the Riemannian gradients of f on M and $\bar{f}$ on $\bar{M}$ satisfy*

$$\operatorname{grad} f(x) = P_x(\operatorname{grad} \bar{f}(x)), \quad x \in M \subset \bar{M}.$$

Proof Let $\zeta_x := P_x(\operatorname{grad} \bar{f}(x)) \in T_x M$ and $\eta_x := \operatorname{grad} \bar{f}(x) - \zeta_x \in N_x M$. For any $\xi \in T_x M$, we have

$$g_x(\zeta_x, \xi) = \bar{g}_x(\zeta_x + \eta_x, \xi) = \bar{g}_x(\operatorname{grad} \bar{f}(x), \xi) = D\bar{f}(x)[\xi] = Df(x)[\xi],$$

where we identify $T_x M$ and $N_x M$ with subspaces of $T_x \bar{M}$ as in Remark 3.4. It follows from Definition 3.15 that $\operatorname{grad} f(x) = \zeta_x$. This completes the proof. $\square$

Example 3.9 The Stiefel manifold $\operatorname{St}(p, n)$ with the Riemannian metric (3.12) is a Riemannian submanifold of $\mathbb{R}^{n \times p}$, as discussed in Example 3.7. We derive the orthogonal projection P_X onto the tangent space $T_X \operatorname{St}(p, n)$ at arbitrary $X \in \operatorname{St}(p, n)$.

We first discuss the normal space $N_X \operatorname{St}(p, n) = (T_X \operatorname{St}(p, n))^{\perp}$ using an arbitrarily fixed $X_{\perp}$ satisfying $X_{\perp}^T X_{\perp} = I_{n-p}$ and $X^T X_{\perp} = 0$. The discussion on deriving (3.10) implies that any $V \in N_X \operatorname{St}(p, n) \subset \mathbb{R}^{n \times p}$ can be expressed as $V = XV_1 + X_{\perp} V_2$ with some $V_1 \in \mathbb{R}^{p \times p}$ and $V_2 \in \mathbb{R}^{(n-p) \times p}$. As $V \in N_X \operatorname{St}(p, n)$, we have $\operatorname{tr}(V^T \xi) = 0$ for any $\xi = XB + X_{\perp} C \in T_X \operatorname{St}(p, n)$ with $B \in \operatorname{Skew}(p)$ and $C \in \mathbb{R}^{(n-p) \times p}$. Consequently, we have

$$0 = \text{tr}(V^T \xi) = \text{tr}((XV_1 + X_\perp V_2)^T (XB + X_\perp C)) = \text{tr}(V_1^T B + V_2^T C)$$

for any $B \in \text{Skew}(p)$ and $C \in \mathbb{R}^{(n-p) \times p}$, which is equivalent to the condition that V_1 is symmetric and $V_2 = 0$. Therefore, the normal space is written as

$$N_X \text{St}(p, n) = \{XS \mid S \in \text{Sym}(p)\}.$$

We proceed to the orthogonal projection P_X. Any $Y \in \mathbb{R}^{n \times p}$ is decomposed into $Y = U + V$ with some $U \in T_X \text{St}(p, n)$ and $V \in N_X \text{St}(p, n)$. It follows that there exist $B \in \text{Skew}(p)$, $C \in \mathbb{R}^{(n-p) \times p}$, and $S \in \text{Sym}(p)$ such that $U = XB + X_\perp C$ and $V = XS$. Multiplying $Y = U + V$ by X and $X_\perp$ from the left, we obtain $X^T Y = B + S$ and $X_\perp^T Y = C$, respectively. Since $B \in \text{Skew}(p)$ and $S \in \text{Sym}(p)$, they are the skew-symmetric and symmetric parts of $X^T Y$, respectively, i.e., $B = \text{skew}(X^T Y)$ and $S = \text{sym}(X^T Y)$. Furthermore, $(X, X_\perp) \in \mathcal{O}_n$ implies $(X, X_\perp)(X, X_\perp)^T = I_n$, which is written as $XX^T + X_\perp X_\perp^T = I_n$. By using these relations, P_X acts on $Y \in \mathbb{R}^{n \times p}$ as

$$\begin{aligned} P_X(Y) &= U = XB + X_\perp C = X \text{skew}(X^T Y) + X_\perp X_\perp^T Y \\ &= X \text{skew}(X^T Y) + (I_n - XX^T)Y = Y - X \text{sym}(X^T Y). \end{aligned} \tag{3.19}$$

Combining the above discussion with Proposition 3.3, for a smooth function $\bar{f} : \mathbb{R}^{n \times p} \to \mathbb{R}$ whose restriction to $\text{St}(p, n)$ is $f : \text{St}(p, n) \to \mathbb{R}$, its Euclidean gradient $\nabla \bar{f}(X)$ at $X \in \text{St}(p, n)$ can be used to compute the Riemannian gradient $\text{grad } f(X)$ of f as

$$\text{grad } f(X) = P_X(\nabla \bar{f}(X)) = \nabla \bar{f}(X) - X \text{sym}(X^T \nabla \bar{f}(X)).$$

For some concrete examples for f, see Sect. 5.1.

The sphere S^{n-1} is considered a special case of the Stiefel manifold, i.e., $S^{n-1} = \text{St}(1, n)$. Therefore, the gradient of $f : S^{n-1} \to \mathbb{R}$ at $x \in S^{n-1}$ is

$$\text{grad } f(x) = P_x(\nabla \bar{f}(x)) = (I_n - xx^T)\nabla \bar{f}(x).$$

The special case of $n = 3$ was discussed as (1.5) in Sect. 1.1.2. When $\bar{f}(x) := c^T x$ and $f := \bar{f}|_{S^{n-1}}$, we have $\nabla \bar{f}(x) = c$ and can again obtain the result in Example 3.8.

The gradient is an important tool for optimization algorithms. Furthermore, the gradient gives the *first-order necessary optimality condition* for Problem 3.1.

Theorem 3.4 *If f is of class C^1 and $x_* \in M$ is a local optimal solution to Problem 3.1, then we have* $\text{grad } f(x_*) = 0$.

Proof For an arbitrary $\xi \in T_{x_*} M$, let γ be a curve on M satisfying $\gamma(0) = x_*$ and $\dot{\gamma}(0) = \xi$. We define the one variable function $g : \mathbb{R} \to \mathbb{R}$ as $g(t) := f(\gamma(t))$. As x_* is a local optimal solution, $t = 0$ is a local optimal solution to the minimization problem of g. Therefore, Theorem 2.2 with $n = 1$ yields $g'(0) = 0$. Because we have $g'(0) = Df(\gamma(0))[\dot\gamma(0)] = \langle \operatorname{grad} f(x_*), \xi \rangle_{x_*}$, this means that $\langle \operatorname{grad} f(x_*), \xi \rangle_{x_*} = 0$. Since ξ is arbitrary, we obtain $\operatorname{grad} f(x_*) = 0$, which is our claim. □

This theorem provides a guideline for development of Riemannian optimization algorithms, which is similar to that for the Euclidean case, i.e., we seek a critical point $x_* \in M$ of f, at which $\operatorname{grad} f(x_*) = 0$.

3.4 Retractions

In line search in the Euclidean space $\mathbb{R}^n$, a step length is chosen as a minimizer (or its approximation) of the one variable function (2.19). On a Riemannian manifold M, however, we cannot define a line in general; therefore, we use a curve on M to perform a similar search. In $\mathbb{R}^n$, the line $l_k(t) := x_k + td_k$ emanating from $x_k \in \mathbb{R}^n$ in the direction of $d_k \in \mathbb{R}^n$ satisfies $l_k(0) = x_k$ and $l'_k(0) = d_k$. Similarly, on M, we use a curve γ_k emanating from $x_k \in M$ in the direction of $\eta_k \in T_{x_k}M$ such that

$$\gamma_k(0) = x_k, \qquad \dot\gamma_k(0) = \eta_k. \tag{3.20}$$

In Sect. 1.1.2, we discussed the curve $\gamma_k(t) := R_{x_k}(t\eta_k)$ by a map $R_{x_k}: T_{x_k}M \to M$ for $M = S^2$. Fixing $x_k \in M$ arbitrarily, for any $\eta_k \in T_{x_k}M$, the condition (3.20) is equivalent to

$$R_{x_k}(0) = x_k, \qquad DR_{x_k}(0)[\eta_k] = \eta_k.$$

Since x_k is arbitrary, extending the point to the whole M to obtain a map R and rewriting the above conditions yield the following definition of a retraction.[2]

Definition 3.16 A smooth map $R: TM \to M$ is called a *retraction* on a smooth manifold M if the restriction of R to the tangent space T_xM at any point $x \in M$, denoted by R_x, satisfies the following conditions:

1. $R_x(0_x) = x$.
2. $DR_x(0_x) = \operatorname{id}_{T_xM}$.

Figure 3.3 is a conceptual illustration of a retraction. The concept of a retraction was proposed by Shub (1986) and by Adler et al. (2002). Definition 3.16 is adopted from Absil et al. (2008).

Example 3.10 We introduce three types of retractions on the sphere S^{n-1} endowed with the induced metric (3.11). The maps R^1 and R^2, defined as

---

[2]The concept of a retraction here is not the same as a retraction discussed in topology (see, e.g., Lee (2012)).

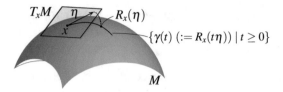

Fig. 3.3 Conceptual illustration of retraction R. The retraction R maps a tangent vector η at a point x to another point on M. Then, the curve defined by $\gamma(t) := R_x(t\eta)$ satisfies $\gamma(0) = x$ and $\dot{\gamma}(0) = \eta$

$$R_x^1(\eta) := (\cos\|\eta\|_x)x + \frac{\sin\|\eta\|_x}{\|\eta\|_x}\eta, \qquad R_x^2(\eta) := \frac{x+\eta}{\|x+\eta\|_2} = \frac{x+\eta}{\sqrt{1+\|\eta\|_x^2}}$$

$$(3.21)$$

for $x \in S^{n-1}$ and $\eta \in T_x S^{n-1}$, are the examples of retractions we introduced for the case of $n = 2$ in Sect. 1.1.2, where we define $R_x^1(0) := x$. Note that R^1 is guaranteed to be smooth by defining $R_x^1(\eta)$ as (3.21) for $\eta \neq 0$ and $R_x^1(0) := x$. Here, R^1 is a map called the *exponential map* on S^{n-1}. In addition, R^3 defined as

$$R_x^3(\eta) := \sqrt{1 - \|\eta\|_x^2}\, x + \eta, \quad \|\eta\|_x < 1 \qquad (3.22)$$

satisfies the two conditions in Definition 3.16. However, we note that R^3 is not defined over the whole TM. This R^3 is called an *orthographic retraction* (Absil and Malick 2012).

Fixing $x \in M$ and $\eta \neq 0$ on $T_x M$ arbitrarily, the image of $\{t\eta \mid t \geq 0\}$ by R^2 is a proper subset of that by R^1; therefore, they are not equal. Further, we note that $R^3(t\eta)$ is defined only when $0 \leq t < 1/\|\eta\|_x$.

Example 3.11 For the Stiefel manifold $\mathrm{St}(p, n)$, consider $R \colon T\,\mathrm{St}(p, n) \to \mathrm{St}(p, n)$ defined by

$$R_X(\eta) = \mathrm{qf}(X + \eta) \qquad (3.23)$$

for an arbitrary point $X \in \mathrm{St}(p, n)$ and tangent vector $\eta \in T_X\,\mathrm{St}(p, n)$, where qf returns the Q-factor of the QR decomposition of the matrix in parentheses (Absil et al. 2008). Then, R is a retraction on $\mathrm{St}(p, n)$. To prove this, we verify that the two conditions in Definition 3.16 are satisfied. Regarding the first condition, we have $R_X(0) = \mathrm{qf}(X) = X$ because $X = XI_p$ is the QR decomposition of $X \in \mathrm{St}(p, n)$. The second condition can be verified as $DR_X(0)[\eta] = D\,\mathrm{qf}(X)[\eta] = \eta$. Here, the second equality follows from Proposition 3.4 and Corollary 3.1 below (see also Absil et al. (2008)). Thus, R defined by (3.23) is a retraction on $\mathrm{St}(p, n)$. In the case of $p = 1$, this R reduces to R^2 in Example 3.10.

Proposition 3.4 *Let n and p be positive integers satisfying $p \leq n$ and $\mathbb{R}_*^{n \times p}$ be the set of all $n \times p$ matrices of rank p, i.e., full-rank. Let $\mathrm{qf} \colon \mathbb{R}_*^{n \times p} \to \mathrm{St}(p, n)$ be a map that returns the Q-factor of the QR decomposition of the input. Then, we have*

$$\mathrm{D}\,\mathrm{qf}(X)[Y] = Q\rho_{\mathrm{skew}}(Q^T Y R^{-1}) + (I_n - QQ^T)Y R^{-1}, \qquad (3.24)$$

where $X \in \mathbb{R}_*^{n \times p}$, $Y \in \mathbb{R}^{n \times p}$, $Q := \mathrm{qf}(X) \in \mathrm{St}(p, n)$, $R := Q^T X$, i.e., $X = QR$ is the QR decomposition of X, and ρ_{skew} acts on a square matrix $A = (a_{ij})$ as $\rho_{\mathrm{skew}}(A) = (r_{ij})$, where $r_{ij} = a_{ij}$ if $i > j$, $r_{ij} = 0$ if $i = j$, and $r_{ij} = -a_{ji}$ if $i < j$.

Proof Let $X + tY = Q(t)R(t)$ be the QR decomposition of $X + tY$, where $Q(t) = \mathrm{qf}(X + tY)$. Note that $Q(0) = Q$, $R(0) = R$, and $\dot{Q}(0) = \mathrm{D}\,\mathrm{qf}(X)[Y]$. Thus, our goal is to find an explicit formula for $\dot{Q}(0)$. We write $\dot{Q} := \dot{Q}(0)$ and $\dot{R} := \dot{R}(0)$.

Remark 2.1 implies that $Q(t)$ and $R(t)$ are differentiable. Differentiating $X + tY = Q(t)R(t)$ at $t = 0$, we obtain

$$Y = \dot{Q}R + Q\dot{R}. \qquad (3.25)$$

Hence, we obtain $Q^T \dot{Q} = Q^T Y R^{-1} - \dot{R}R^{-1}$. Here, because $-\dot{R}R^{-1}$ is an upper triangular matrix, the lower triangular part (excluding the diagonal part) of $Q^T \dot{Q}$ coincides with that of $Q^T Y R^{-1}$. Furthermore, as $\dot{Q} \in T_Q \mathrm{St}(p, n)$, the expression (3.8) yields $Q^T \dot{Q} \in \mathrm{Skew}(p)$. Therefore, $Q^T \dot{Q} = \rho_{\mathrm{skew}}(Q^T Y R^{-1})$. In addition, we multiply (3.25) by $(I_n - QQ^T)$ and R^{-1} from the left and right, respectively, to obtain $(I_n - QQ^T)\dot{Q} = (I_n - QQ^T)Y R^{-1}$. As a result, we have

$$\begin{aligned}
\mathrm{D}\,\mathrm{qf}(X)[Y] = \dot{Q} &= QQ^T \dot{Q} + (I_n - QQ^T)\dot{Q} \\
&= Q\rho_{\mathrm{skew}}(Q^T Y R^{-1}) + (I_n - QQ^T)Y R^{-1}.
\end{aligned}$$

This completes the proof. $\square$

Corollary 3.1 Under the assumptions of Proposition 3.4, if $X \in \mathrm{St}(p, n)$ and $Y \in T_X \mathrm{St}(p, n)$, then we have $\mathrm{D}\,\mathrm{qf}(X)[Y] = Y$.

Proof We have the QR decomposition of $X \in \mathrm{St}(p, n)$ as $X = QR$ with $Q = X$ and $R = I_p$. Further, for $Y \in T_X \mathrm{St}(p, n)$, it follows from (3.8) that $X^T Y$ is skew-symmetric, implying $\rho_{\mathrm{skew}}(Q^T Y R^{-1}) = \rho_{\mathrm{skew}}(X^T Y) = X^T Y$. Then, the desired conclusion straightforwardly follows from (3.24). $\square$

For two manifolds M and N that are diffeomorphic to each other, we have the following theorem (Sato and Aihara 2019), which can be used to construct a retraction on a manifold M from an already known retraction on N.

Theorem 3.5 Let M and N be smooth manifolds that are diffeomorphic to each other and $\Phi: M \to N$ be a diffeomorphism. When R^N is a retraction on N, R^M defined by

$$R_x^M(\eta) := \Phi^{-1}(R_{\Phi(x)}^N(\mathrm{D}\Phi(x)[\eta])), \quad \eta \in T_x M, \ x \in M \qquad (3.26)$$

is a retraction on M.

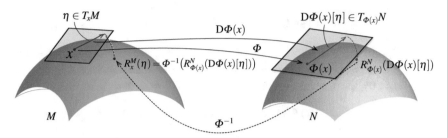

Fig. 3.4 Based on a given retraction R^N on N, a new retraction R^M on M is constructed, where M and N are manifolds diffeomorphic to each other

Proof In this proof, to avoid confusion, let 0_x and $0_{\Phi(x)}$ denote the zero vectors in $T_x M$ and $T_{\Phi(x)} N$, respectively. It is enough to check that R^M defined by (3.26) satisfies the two conditions in Definition 3.16. As $D\Phi(x)\colon T_x M \to T_{\Phi(x)} N$ is a linear map, we have $D\Phi(x)[0_x] = 0_{\Phi(x)}$. The first condition can then be easily checked, as

$$R_x^M(0_x) = \Phi^{-1}(R_{\Phi(x)}^N(D\Phi(x)[0_x])) = \Phi^{-1}(R_{\Phi(x)}^N(0_{\Phi(x)})) = \Phi^{-1}(\Phi(x)) = x,$$

where the third equality follows from the first condition in Definition 3.16 for R^N. To prove that the second condition in Definition 3.16 holds for R^M, we calculate $DR_x^M(0_x)[\eta]$ for arbitrary $\eta \in T_x M$. Letting γ be a curve on N defined as $\gamma(t) := R_{\Phi(x)}^N(t D\Phi(x)[\eta])$, we have $\gamma(0) = \Phi(x)$ and $\dot{\gamma}(0) = D\Phi(x)[\eta]$ since R^N is a retraction on N. Then, we have $R_x^M(t\eta) = \Phi^{-1}(\gamma(t))$ and

$$\begin{aligned} DR_x^M(0_x)[\eta] &= D\Phi^{-1}(\gamma(0))[\dot{\gamma}(0)] = D\Phi^{-1}(\Phi(x))[D\Phi(x)[\eta]] \\ &= D\left(\Phi^{-1} \circ \Phi\right)(x)[\eta] = D\,\mathrm{id}(x)[\eta] = \eta, \end{aligned}$$

where the first and third equalities follow from Proposition 3.2.   □

Figure 3.4 shows the scenario described in Theorem 3.5.

The following example shows one application of Theorem 3.5.

Example 3.12   Consider the Stiefel manifold $\mathrm{St}(p, n)$ and *generalized Stiefel manifold* $\mathrm{St}_G(p, n)$, where $G \in \mathrm{Sym}(n)$ is a positive definite matrix and

$$\mathrm{St}_G(p, n) := \left\{ X \in \mathbb{R}^{n \times p} \mid X^T G X = I_p \right\}.$$

The condition $X \in \mathrm{St}_G(p, n)$ is equivalent to $\left(\sqrt{G} X\right)^T \left(\sqrt{G} X\right) = I_p$, implying $\sqrt{G} X \in \mathrm{St}(p, n)$. Thus, we can define a diffeomorphism $\Phi \colon \mathrm{St}_G(p, n) \to \mathrm{St}(p, n)$ as

$$\Phi(X) = \sqrt{G} X, \quad X \in \mathrm{St}_G(p, n).$$

Recall the QR-based retraction R on the Stiefel manifold $\mathrm{St}(p, n)$ defined by (3.23). Then, it follows from Theorem 3.5 that $R^G \colon T\,\mathrm{St}_G(p, n) \to \mathrm{St}_G(p, n)$ defined by

$$R_X^G(\eta) := \Phi^{-1}(R_{\Phi(X)}(D\Phi(X)[\eta])), \quad \eta \in T_X \operatorname{St}_G(p, n), \ X \in \operatorname{St}_G(p, n)$$

is a retraction on $\operatorname{St}_G(p, n)$. We can further calculate the right-hand side as

$$R_X^G(\eta) = \sqrt{G}^{-1} \operatorname{qf}\left(\sqrt{G}(X + \eta)\right).$$

We apply this retraction to the canonical correlation analysis in Sect. 5.3.1.

3.5 Step Lengths

In Sect. 2.9.1, we introduced the concept of step length selection in the Euclidean space $\mathbb{R}^n$ on the basis of exact line search and the Armijo, Wolfe, and strong Wolfe conditions. Here, we extend these concepts to Riemannian manifolds.

Note that all choices in Sect. 2.9.1 were made by observing how the value of the objective function varies on the half-line $\{x_k + td_k \mid t > 0\}$, as in (2.19). Let $(M, \langle \cdot, \cdot \rangle)$ be a Riemannian manifold with a retraction R. Assume that the current point $x_k \in M$ and *search direction*[3] $\eta_k \in T_{x_k} M$ are already given and that η_k satisfies

$$\langle \operatorname{grad} f(x_k), \eta_k \rangle_{x_k} < 0. \tag{3.27}$$

To perform a search on the curve $\gamma_k(t) := R_{x_k}(t\eta_k)$ instead of a half-line, we define the function ϕ_k of one variable as

$$\phi_k(t) := f\left(R_{x_k}(t\eta_k)\right), \tag{3.28}$$

instead of (2.19). Then, we have

$$\phi_k'(t) = Df\left(R_{x_k}(t\eta_k)\right)\left[DR_{x_k}(t\eta_k)[\eta_k]\right] = \left\langle \operatorname{grad} f\left(R_{x_k}(t\eta_k)\right), DR_{x_k}(t\eta_k)[\eta_k]\right\rangle_{R_{x_k}(t\eta_k)}$$

to obtain $\phi_k'(0) = \langle \operatorname{grad} f(x_k), \eta_k \rangle_{x_k}$ by the definition of a retraction. As the condition (3.27) can be written as $\phi_k'(0) < 0$, we call η_k satisfying this condition a *descent direction* of f at x_k. It is then natural to define the step length t_k through exact search as

$$t_k := \arg\min_{t \in T_k} f\left(R_{x_k}(t\eta_k)\right),$$

where T_k is the set of all $t > 0$ for which $R_{x_k}(t\eta_k)$ is defined. Note that $R_{x_k}(t\eta_k)$ is not necessarily defined for all $t \geq 0$, as indicated by the example (3.22).

The conditions (2.23), (2.25), and (2.27), on which the Armijo and (strong) Wolfe conditions are based, can also be generalized to M, by considering the conditions

---

[3] Here, η_k corresponds to d_k in Sect. 2.9.1. We use the symbol η in the Riemannian case to emphasize that it is a tangent vector.

(2.22), (2.24), and (2.26), in which ϕ_k is replaced with that defined in (3.28). In other words, for constants c_1 and c_2 with $0 < c_1 < c_2 < 1$, the *Armijo condition* on M is

$$f\left(R_{x_k}(t_k\eta_k)\right) \leq f(x_k) + c_1 t_k \langle \operatorname{grad} f(x_k), \eta_k \rangle_{x_k}, \tag{3.29}$$

the *Wolfe conditions* are the set of (3.29) and

$$\left\langle \operatorname{grad} f\left(R_{x_k}(t_k\eta_k)\right), \operatorname{D}R_{x_k}(t_k\eta_k)[\eta_k] \right\rangle_{R_{x_k}(t_k\eta_k)} \geq c_2 \langle \operatorname{grad} f(x_k), \eta_k \rangle_{x_k}, \tag{3.30}$$

and the *strong Wolfe conditions* are the set of (3.29) and

$$\left| \left\langle \operatorname{grad} f\left(R_{x_k}(t_k\eta_k)\right), \operatorname{D}R_{x_k}(t_k\eta_k)[\eta_k] \right\rangle_{R_{x_k}(t_k\eta_k)} \right| \leq c_2 \left| \langle \operatorname{grad} f(x_k), \eta_k \rangle_{x_k} \right|. \tag{3.31}$$

The existence of a step length that satisfies the Armijo or (strong) Wolfe conditions can be shown by an analysis similar to that for the Euclidean case (see, e.g., Nocedal and Wright (2006)). Here, we prove this for a manifold case. For the theoretical existence of a step length satisfying these conditions, it is sufficient to give a proof for the strong Wolfe conditions, as the Armijo and Wolfe conditions are milder.

Proposition 3.5 *Let f be a C^1 real-valued function defined on a smooth Riemannian manifold $(M, \langle \cdot, \cdot \rangle)$ with a retraction R, and assume that R is defined over the whole TM. Further, assume that a point $x_k \in M$ and a descent direction $\eta_k \in T_{x_k}M$ of f at x_k are given. If f is lower-bounded on the curve $\{R_{x_k}(t\eta_k) \mid t > 0\}$ and constants c_1 and c_2 satisfy $0 < c_1 < c_2 < 1$, then there exists a step length $t_k > 0$ satisfying the strong Wolfe conditions (3.29) and (3.31).*

Proof We define ϕ_k as (3.28). First, we prove the existence of t_k satisfying (3.29). Because $\eta_k \in T_{x_k}M$ is a descent direction, we have $\phi_k'(0) < 0$. From the assumption, ϕ_k is lower-bounded on $(0, \infty)$, whereas $c_1\phi_k'(0)t + \phi_k(0)$ can be any small number by taking a sufficiently large t; therefore, noting the relation $\phi_k'(0) < c_1\phi_k'(0) < 0$, there exists $t > 0$ such that $\phi_k(t) = c_1\phi_k'(0)t + \phi_k(0)$; i.e., the curve $y = \phi_k(t)$ and line $y = c_1\phi_k'(0)t + \phi_k(0)$ on the t–y plane intersect at some t. Let α_1 be the smallest number among such t. Then, $\phi_k'(0) < c_1\phi_k'(0) < 0$ implies

$$\phi_k(t) < c_1\phi_k'(0)t + \phi_k(0)$$

for $0 < t < \alpha_1$ and

$$\phi_k(\alpha_1) = c_1\phi_k'(0)\alpha_1 + \phi_k(0). \tag{3.32}$$

Thus, any $t_k \in (0, \alpha_1]$ satisfies (3.29).

From the mean value theorem for ϕ_k together with (3.32), there exists $\alpha_2 \in (0, \alpha_1)$ such that

$$\phi_k'(\alpha_2) = \frac{\phi_k(\alpha_1) - \phi_k(0)}{\alpha_1} = c_1\phi_k'(0).$$

It follows from $\phi'_k(0) < 0$ and $0 < c_1 < c_2 < 1$ that $c_2\phi'_k(0) < \phi'_k(\alpha_2) < 0$; this implies that $t_k = \alpha_2 \, (< \alpha_1)$ satisfies (3.29) and (3.31). $\square$

Because the Armijo condition holds for a sufficiently small step length, it is natural to find a step length satisfying the condition as in Algorithm 3.1.

Algorithm 3.1 Backtracking algorithm to find step length satisfying Armijo condition (3.29) on Riemannian manifold M

Input: Riemannian manifold M, retraction R, objective function f on M, current point $x_k \in M$, search direction $\eta_k \in T_{x_k} M$, initial guess $t^{(0)}$, and parameters $c_1, \tau \in (0, 1)$.
Output: Step length $t_k > 0$ satisfying the Armijo condition (3.29).
1: Set $t := t^{(0)}$.
2: **while** $f\left(R_{x_k}(t\eta_k)\right) > f(x_k) + c_1 t \langle \operatorname{grad} f(x_k), \eta_k \rangle_{x_k}$ holds **do**
3:    Set $t := \tau t$.
4: **end while**
5: Set $t_k := t$.

Algorithm 3.1 cannot find a step length satisfying the Wolfe conditions in general. To find a step length for the (strong) Wolfe conditions in the Euclidean space $\mathbb{R}^n$, a modified approach is described in Lemaréchal (1981), Nocedal and Wright (2006). The same approach can be employed for the Riemannian case (Sato 2016).

Once a step length t_k is obtained, we can compute the next point $x_{k+1} \in M$ as

$$x_{k+1} = R_{x_k}(t_k \eta_k). \tag{3.33}$$

Line search methods on M employ this process, which iteratively finds an appropriate step length and updates the current point by (3.33), as in Algorithm 3.2.

Algorithm 3.2 Line search method on Riemannian manifold M

Input: Riemannian manifold M, retraction R, objective function f on M, and initial point $x_0 \in M$.
Output: Sequence $\{x_k\} \subset M$.
1: **for** $k = 0, 1, 2, \ldots$ **do**
2:    Compute a search direction $\eta_k \in T_{x_k} M$.
3:    Compute a step length $t_k > 0$.
4:    Compute the next point as $x_{k+1} = R_{x_k}(t_k \eta_k)$.
5: **end for**

If $M = \mathbb{R}^n$ and we define R as $R_x(d) = x + d$ for $x, d \in \mathbb{R}^n$, this R is a retraction on $\mathbb{R}^n$. Therefore, the above discussion includes the Euclidean case as a special case.

3.6 Riemannian Steepest Descent Method

In Sect. 2.9.3, we discussed the steepest descent method in Euclidean spaces, which is one of the simplest line search methods. This method is easily generalized to the Riemannian case, which we call the *Riemannian steepest descent method*. Specifically, on a Riemannian manifold $(M, \langle \cdot, \cdot \rangle)$ with a retraction R, the search direction at $x_k \in M$ is computed as $\eta_k = -\operatorname{grad} f(x_k) \in T_{x_k} M$. Then, we can apply the updating formula (3.33). The algorithm is described as follows.

Algorithm 3.3 Riemannian steepest descent method

Input: Riemannian manifold M, retraction R, objective function f on M, and initial point $x_0 \in M$.
Output: Sequence $\{x_k\} \subset M$.
1: **for** $k = 0, 1, 2, \ldots$ **do**
2:     Compute a search direction as $\eta_k = -\operatorname{grad} f(x_k)$.
3:     Compute a step length $t_k > 0$.
4:     Compute the next point as $x_{k+1} = R_{x_k}(t_k \eta_k)$.
5: **end for**

Readers may refer to Absil et al. (2008) for further details of the Riemannian steepest descent method. Earlier works using geodesics, i.e., using the exponential map as a retraction, include Edelman et al. (1998), Gabay (1982), Rapcsák (1997), Smith (1994).

3.6.1 Zoutendijk's Theorem

Zoutendijk's theorem is originally related to a series associated with search directions in $\mathbb{R}^n$ (Nocedal and Wright 2006). This theorem is useful for convergence analysis for not only the steepest descent method but also general descent algorithms, such as the conjugate gradient method discussed in the next chapter.

Here, we discuss a Riemannian counterpart to Zoutendijk's theorem to prove a convergence property of the steepest descent method (see also Ring and Wirth (2012), Sato and Iwai (2015)). To describe the claim, we define the angle θ_k between the steepest descent direction $-\operatorname{grad} f(x_k)$ and the search direction $\eta_k \in T_{x_k} M$ at $x_k \in M$. By applying (2.5) to the normed space $T_{x_k} M$ whose norm is induced from $\langle \cdot, \cdot \rangle_{x_k}$, θ_k specifically satisfies

$$\cos \theta_k = \frac{\langle -\operatorname{grad} f(x_k), \eta_k \rangle_{x_k}}{\| -\operatorname{grad} f(x_k) \|_{x_k} \|\eta_k\|_{x_k}} = -\frac{\langle \operatorname{grad} f(x_k), \eta_k \rangle_{x_k}}{\|\operatorname{grad} f(x_k)\|_{x_k} \|\eta_k\|_{x_k}}. \tag{3.34}$$

Zoutendijk's theorem on M is stated as follows.

Theorem 3.6 *Suppose that, in Algorithm 3.2 on a smooth Riemannian manifold* $(M, \langle \cdot, \cdot \rangle)$, *the search direction* η_k *is a descent direction and the step length* t_k *satisfies* *the Wolfe conditions (3.29) and (3.30) for every integer $k \geq 0$. If the objective function f is bounded below and is of class C^1, and if there exists a constant $L > 0$ such that*

$$|D(f \circ R_x)(t\eta)[\eta] - D(f \circ R_x)(0)[\eta]| \leq Lt,$$
$$\eta \in T_x M \text{ with } \|\eta\|_x = 1, \ x \in M, \ t \geq 0, \tag{3.35}$$

then, for θ_k defined by (3.34), the following series converges:

$$\sum_{k=0}^{\infty} \cos^2 \theta_k \|\operatorname{grad} f(x_k)\|_{x_k}^2 < \infty. \tag{3.36}$$

Proof We prove the claim in a manner similar to the Euclidean case in Nocedal and Wright (2006).

From (3.30) and (3.35), we have

$$(c_2 - 1)\langle \operatorname{grad} f(x_k), \eta_k \rangle_{x_k}$$
$$\leq \langle \operatorname{grad} f(x_{k+1}), DR_{x_k}(t_k \eta_k)[\eta_k] \rangle_{x_{k+1}} - \langle \operatorname{grad} f(x_k), \eta_k \rangle_{x_k}$$
$$= Df(x_{k+1})[DR_{x_k}(t_k \eta_k)[\eta_k]] - Df(x_k)[\eta_k]$$
$$= Df(R_{x_k}(t_k \eta_k))[DR_{x_k}(t_k \eta_k)[\eta_k]] - Df(R_{x_k}(0))[DR_{x_k}(0)[\eta_k]]$$
$$= D(f \circ R_{x_k})(t_k \eta_k)[\eta_k] - D(f \circ R_{x_k})(0)[\eta_k]$$
$$\leq \|\eta_k\|_{x_k} \left| D(f \circ R_{x_k}) \left(t_k \|\eta_k\|_{x_k} \frac{\eta_k}{\|\eta_k\|_{x_k}} \right) \left[\frac{\eta_k}{\|\eta_k\|_{x_k}} \right] - D(f \circ R_{x_k})(0) \left[\frac{\eta_k}{\|\eta_k\|_{x_k}} \right] \right|$$
$$\leq Lt_k \|\eta_k\|_{x_k}^2.$$

Therefore, we have

$$t_k \geq -\frac{(1 - c_2)\langle \operatorname{grad} f(x_k), \eta_k \rangle_{x_k}}{L\|\eta_k\|_{x_k}^2},$$

which, together with the Armijo condition (3.29) and $\langle \operatorname{grad} f(x_k), \eta_k \rangle_{x_k} < 0$, yields

$$f(x_{k+1}) \leq f(x_k) + c_1 t_k \langle \operatorname{grad} f(x_k), \eta_k \rangle_{x_k}$$
$$\leq f(x_k) - \frac{c_1(1 - c_2)\langle \operatorname{grad} f(x_k), \eta_k \rangle_{x_k}^2}{L\|\eta_k\|_{x_k}^2}$$
$$= f(x_k) - \frac{c_1(1 - c_2)}{L} \cos^2 \theta_k \|\operatorname{grad} f(x_k)\|_{x_k}^2$$
$$\leq f(x_0) - \frac{c_1(1 - c_2)}{L} \sum_{j=0}^{k} \cos^2 \theta_j \|\operatorname{grad} f(x_j)\|_{x_j}^2,$$

where the equality follows from (3.34). Because f is bounded below, there exists a constant C such that $f(x_0) - f(x_{k+1}) \leq C$ for all $k \geq 0$. Then, we have

$$\sum_{j=0}^{k} \cos^2 \theta_j \|\mathrm{grad}\, f(x_j)\|_{x_j}^2 \leq \frac{L}{c_1(1 - c_2)}(f(x_0) - f(x_{k+1})) \leq \frac{LC}{c_1(1 - c_2)}. \tag{3.37}$$

As the right-hand side is a constant, taking the limit $k \to \infty$ in (3.37) yields the desired property (3.36). This completes the proof. $\square$

Remark 3.5 Inequality (3.35) is a weaker condition than the Lipschitz continuous differentiability of $f \circ R_x$. It is shown in Sato and Iwai (2015) that Eq. (3.35) holds for objective functions in some practical Riemannian optimization problems. The relation with the standard Lipschitz continuous differentiability is also discussed further in Sato and Iwai (2015).

3.6.2 Global Convergence Analysis

When (3.35) is satisfied, the global convergence of the steepest descent method with the Wolfe conditions can easily be shown by using the Riemannian version of Zoutendijk's theorem.

Theorem 3.7 *Assume that f is bounded below and of class C^1 and that (3.35) holds for some $L > 0$. Then, the sequence $\{x_k\}$ generated by Algorithm 3.3, where the step length t_k satisfies the Wolfe conditions (3.29) and (3.30) for every integer $k \geq 0$, satisfies*

$$\lim_{k \to \infty} \|\mathrm{grad}\, f(x_k)\|_{x_k} = 0.$$

Proof Let θ_k be the angle between $-\mathrm{grad}\, f(x_k)$ and $\eta_k := -\mathrm{grad}\, f(x_k)$. Then, we have $\cos \theta_k = 1$ for all k. It follows from Theorem 3.6 that $\sum_{k=0}^{\infty} \|\mathrm{grad}\, f(x_k)\|_{x_k}^2 < \infty$. Hence, $\|\mathrm{grad}\, f(x_k)\|_{x_k}$ must converge to 0 as $k \to \infty$. This ends the proof. $\square$

This proof provides insight into a more general case in which we do not necessarily choose $-\mathrm{grad}\, f(x_k)$ as the search direction η_k. Let θ_k be the angle between $-\mathrm{grad}\, f(x_k)$ and η_k. Whenever $\cos \theta_k$ is bounded away from 0, a discussion similar to that in the above proof yields $\lim_{k \to \infty} \|\mathrm{grad}\, f(x_k)\|_{x_k} = 0$.

Chapter 4
Conjugate Gradient Methods on Riemannian Manifolds

In this chapter, we discuss the conjugate gradient (CG) methods on Riemannian manifolds, which we also call Riemannian CG methods. They can be considered to be a modified version of the Riemannian steepest descent method. However, to obtain further insights, we first review the CG methods in Euclidean spaces; we call these Euclidean CG methods.

Historically, the first CG method was proposed in Hestenes and Stiefel (1952) as an iterative method for solving a system of linear equations by minimizing a closely related quadratic function. This is called the linear CG method, as introduced in Sect. 4.1. It was generalized for optimization methods to minimize more general functions (Fletcher and Reeves 1964). The generalized methods are called the nonlinear CG methods, as discussed in Sect. 4.2. Although the nonlinear CG algorithms are similar to the steepest descent method, they generally exhibit superior performance.

Analogous to the steepest descent method, it is natural to generalize the Euclidean CG methods to those on Riemannian manifolds. In Sects. 4.3–4.6, we discuss the Riemannian CG methods in detail. In particular, we analyze the Fletcher–Reeves-type and Dai–Yuan-type Riemannian CG methods and prove their global convergence properties under some conditions.

4.1 Linear Conjugate Gradient Method

In this section, we address systems of linear equations, or linear systems, which are written as

$$Ax = b, \tag{4.1}$$

where $A \in \mathbb{R}^{n \times n}$ and $b \in \mathbb{R}^n$ are constants and $x \in \mathbb{R}^n$ is a vector of unknown variables. Such linear systems appear in every field and solving them has been an impor-

© Springer Nature Switzerland AG 2021
H. Sato, *Riemannian Optimization and Its Applications*,
SpringerBriefs in Control, Automation and Robotics,
https://doi.org/10.1007/978-3-030-62391-3_4

tant subject of research. If A is invertible, then the solution to (4.1) is theoretically given by $x_* = A^{-1}b$. However, the computational cost of A^{-1} is relatively high and it is rare in a large-scale problem that (4.1) is solved by computing the inverse matrix A^{-1}, followed by multiplying b by A^{-1} from the left. Approaches for solving (4.1)—which include iterative methods, such as the Krylov subspace methods, and direct methods, such as Gaussian elimination—have been studied extensively (Saad 2003).

In this section, both sequences of vectors and elements of vectors appear in the discussion. To distinguish them, we use the $v_i \in \mathbb{R}^n$ notation for the ith vector of a sequence $\{v_i\} \subset \mathbb{R}^n$ and $(v)_i \in \mathbb{R}$ notation for the ith element of a vector $v \in \mathbb{R}^n$.

The *Krylov subspace* of order k generated by $A \in \mathbb{R}^{n \times n}$ and a vector $r_0 \in \mathbb{R}^n$ is defined as $\mathcal{K}_k(A, r_0) := \mathrm{span}\{r_0, Ar_0, A^2r_0, \ldots, A^{k-1}r_0\}$. The *Krylov subspace methods* with a given initial point x_0 for solving (4.1) generate a sequence $\{x_k\}$, where x_k is searched for in $x_0 + \mathcal{K}_k(A, r_0) := \{x_0 + d \mid d \in \mathcal{K}_k(A, r_0)\}$.

In the following, let A be a symmetric positive definite matrix. In this case, an important Krylov subspace method is the *(linear) conjugate gradient method*, which can also be viewed as a method to solve the following unconstrained optimization problem whose objective function $f \colon \mathbb{R}^n \to \mathbb{R}$ is quadratic:

$$\text{minimize} \quad f(x) := \frac{1}{2}x^T Ax - b^T x. \tag{4.2}$$

Because A is symmetric, we can observe in this problem that $\nabla f(x) = Ax - b$ and $\nabla^2 f(x) = A$. This, along with the positive definiteness of A, indicates that $\nabla^2 f(x)$ at each point x is positive definite, implying that f is strictly convex. This fact can also be directly shown according to Example 2.17 and Proposition 2.8. It follows from Proposition 2.10 that the global optimal solution to Problem (4.2) coincides with the critical point of f, i.e., the solution to the linear system (4.1).

Considering the above discussion, we aim to minimize the quadratic function f to solve (4.1). Let us derive an iterative algorithm that generates a sequence $\{x_k\}$ converging to x_*. A sequence $\{x_k\}$ generated by the steepest descent method converges slowly, especially when the condition number of A is large (Nocedal and Wright 2006). Ideally, this issue may be resolved if A can be transformed into a matrix with the minimum condition number 1, i.e., the identity matrix I_n. Theoretically, such a transformation can be achieved by exploiting the Cholesky decomposition of A as

$$A = LL^T, \tag{4.3}$$

where L is an $n \times n$ lower triangular matrix whose diagonal elements are all positive (see Proposition 2.1). The transformation $y = L^T x$ then yields

$$f(x) = \frac{1}{2}x^T LL^T x - b^T L^{-T} L^T x = \frac{1}{2}y^T y - (L^{-1}b)^T y, \tag{4.4}$$

where we can observe that $x^T A x$ is transformed into $y^T y$; i.e., A is transformed into the identity matrix with respect to the new variable y.

Remark 4.1 Using the Cholesky decomposition (4.3), the original linear system $Ax = b$ is transformed into $LL^T x = b$. If (4.3) is numerically practical and L is available, then the solution to the resultant system is obtained by solving $Ly = b$ for y and then solving $L^T x = y$ for x. Both systems can be easily solved by forward and backward substitutions, respectively, noting that L is lower triangular.

It should be noted that the Cholesky decomposition (4.3) may be numerically impractical in large-scale cases. Nevertheless, further discussion in this direction is theoretically worthwhile. Denoting the right-hand side of (4.4) by $g(y)$ and letting $b' := L^{-1}b$, we have

$$g(y) = \sum_{i=1}^{n} \left(\frac{1}{2}(y)_i^2 - (b')_i(y)_i \right), \tag{4.5}$$

implying that we can separately deal with minimization with respect to each $(y)_i$. This is equivalent to sequentially handling optimization for g along the vectors $e_1, e_2, \ldots, e_n$, where $e_i \in \mathbb{R}^n$ is a unit vector whose ith element is 1 and all other elements are 0. In addition, the transformation $y = Pz$ by any $P \in \mathcal{O}_n$ gives

$$g(y) = \frac{1}{2}(Pz)^T(Pz) - b'^T(Pz) = \frac{1}{2}z^T z - (P^T b')^T z = \sum_{i=1}^{n} \left(\frac{1}{2}(z)_i^2 - (b'')_i(z)_i \right), \tag{4.6}$$

where we set $b'' := P^T b'$. The right-hand side of (4.6) has the same form as (4.5), implying that the optimal solution is also obtained by minimization with respect to each $(z)_i$. In other words, the sequence $\{d_0', d_1', \ldots, d_{n-1}'\} \subset \mathbb{R}^n$ of nonzero vectors that are orthogonal to each other with respect to the standard inner product (2.2), which are not necessarily $e_1, e_2, \ldots, e_n$, can generate a sequence $\{y_0, y_1, \ldots, y_n\}$ by $t_k = \arg\min_{t \in \mathbb{R}} g(y_k + t d_k')$ and $y_{k+1} = y_k + t_k d_k'$ for $k = 0, 1, \ldots, n-1$, with y_n being the minimum point of g. Since $P \in \mathcal{O}_n$ in the transformation $y = Pz$ is arbitrary, mutually orthogonal $d_0', d_1', \ldots, d_{n-1}'$ can also be arbitrarily chosen.

The updating formula $y_{k+1} = y_k + t_k d_k'$ with respect to y_k and y_{k+1} in terms of $x_k = L^{-T} y_k$ and $x_{k+1} = L^{-T} y_{k+1}$ implies that $x_{k+1} = x_k + t_k L^{-T} d_k'$. Letting $d_k := L^{-T} d_k'$, the orthogonality condition $(d_k')^T d_l' = 0$ for $k \neq l$ is equivalent to $(L^T d_k)^T (L^T d_l) = 0$, i.e., $d_k^T A d_l = 0$ because of (4.3). Therefore, we endow $\mathbb{R}^n$ with the inner product (see Example 2.1)

$$\langle a, b \rangle_A := a^T A b, \quad a, b \in \mathbb{R}^n, \tag{4.7}$$

and generate a sequence $\{d_0, d_1, \ldots, d_{n-1}\} \subset \mathbb{R}^n$ whose component vectors are nonzero and orthogonal to each other with respect to the inner product $\langle \cdot, \cdot \rangle_A$ (such vectors are said to be *A-conjugate*). Then, we compute $\{x_0, x_1, \ldots, x_n\}$ by

$$t_k = \arg\min_{t \in \mathbb{R}} f(x_k + t d_k), \qquad x_{k+1} = x_k + t_k d_k \qquad (4.8)$$

for $k = 0, 1, \ldots, n - 1$. Since f is quadratic and $d_k^T A d_k > 0$, the step length t_k can be written as (2.21), i.e.,

$$t_k = -\frac{\nabla f(x_k)^T d_k}{d_k^T A d_k}. \qquad (4.9)$$

Note that we then obtain the relation

$$\nabla f(x_{k+1})^T d_k = \frac{d}{dt} f(x_k + t d_k)\bigg|_{t=t_k} = 0. \qquad (4.10)$$

We can modify the steepest descent method and use the negative gradients $-\nabla f(x_0), -\nabla f(x_1), \ldots, -\nabla f(x_{n-1})$ to generate search directions $d_0, d_1, \ldots, d_{n-1}$ that are A-conjugate to each other. Here, we can naturally assume that the n gradients are nonzero because if $\nabla f(x_i) = 0$ holds for some i, then x_i is the optimal solution. If we further assume that they are linearly independent, then the Gram–Schmidt process gives desired search directions satisfying the A-conjugacy as $d_0 = -\nabla f(x_0)$ and

$$d_{k+1} = -\nabla f(x_{k+1}) - \sum_{i=0}^{k} \frac{\langle -\nabla f(x_{k+1}), d_i \rangle_A}{\langle d_i, d_i \rangle_A} d_i, \quad k = 0, 1, \ldots, n - 2, \quad (4.11)$$

where $d_0, d_1, \ldots, d_{n-1}$ are not normalized (see (2.6)). The assumption that the n gradients are linearly independent can be justified by the following proposition.

Proposition 4.1 *Let $A \in \mathrm{Sym}(n)$ be positive definite and consider the optimization problem (4.2). For $\{x_k\}$ and $\{d_k\}$ generated by (4.8), (4.9), and (4.11) with $x_0 \in \mathbb{R}^n$ and $d_0 = -\nabla f(x_0)$, assume that $\nabla f(x_k) \neq 0$ for $k = 0, 1, \ldots, n - 1$. Then, for $k = 1, 2, \ldots, n$, we have*[1]

$$d_l \neq 0, \qquad \nabla f(x_k)^T d_l = \nabla f(x_k)^T \nabla f(x_l) = 0, \quad l = 0, 1, \ldots, k - 1. \qquad (4.12)$$

Proof We prove the statement by induction. For $k = 1$, the claim is evidently valid from $d_0 = -\nabla f(x_0) \neq 0$ and (4.10). We now denote the proposition (4.12) by $P(k)$ and assume that $P(1), P(2), \ldots, P(k)$ are all true for some $k < n$. From the assumption that $\nabla f(x_0), \nabla f(x_1), \ldots, \nabla f(x_k)$ are not 0 and are orthogonal to each other with respect to the standard inner product, as implied by (4.12), they are linearly independent. Then, the Gram–Schmidt process (4.11) yields $d_k \neq 0$ and $\langle d_k, d_l \rangle_A = d_k^T A d_l = 0$ for $l = 0, 1, \ldots, k - 1$. The quantity $\nabla f(x_{k+1})^T d_l$ is equal to 0 for $l = k$ from (4.10) and

---

[1] The statement is equivalent to $d_k \neq 0$ for $k = 0, 1, \ldots, n - 1$ and $\nabla f(x_k)^T d_l = \nabla f(x_k)^T \nabla f(x_l) = 0$ for $k, l = 0, 1, \ldots, n$ with $l < k$. The expression (4.12) is for the ease of proof by induction.

$$\nabla f(x_{k+1})^T d_l = (A(x_k + t_k d_k) - b)^T d_l = \nabla f(x_k)^T d_l + t_k d_k^T A d_l = 0$$

for $l = 0, 1, \ldots, k - 1$ by the induction hypothesis. It follows from (4.11) that

$$\nabla f(x_{k+1})^T \nabla f(x_l) = -\nabla f(x_{k+1})^T \left(d_l + \sum_{i=0}^{l-1} \frac{\langle -\nabla f(x_l), d_i \rangle_A}{\langle d_i, d_i \rangle_A} d_i \right) = 0$$

for $l = 0, 1, \ldots, k$. Thus, $P(k + 1)$ is also true. Therefore, (4.12) holds true for $k = 1, 2, \ldots, n$. $\square$

Consequently, $\{\nabla f(x_0), \nabla f(x_1), \ldots, \nabla f(x_{n-1})\}$ are linearly independent under the assumption $\nabla f(x_k) \neq 0$ for $k = 0, 1, \ldots, n - 1$; therefore, $\{d_0, d_1, \ldots, d_{n-1}\}$ is an orthogonal basis of $\mathbb{R}^n$ with respect to the inner product (4.7). Thus, we do not need to explicitly assume that the n gradients are linearly independent since we have shown that the condition is automatically satisfied. It is important to note in this case that we have $\nabla f(x_n) = 0$, i.e., x_n is the optimal solution to the problem (4.2) because $\{\nabla f(x_0), \nabla f(x_1), \ldots, \nabla f(x_{n-1})\}$ is an orthogonal basis of $\mathbb{R}^n$ with respect to the standard inner product and $\nabla f(x_n)$ is orthogonal to the n gradients.

In addition, (4.11) can be rewritten as a simpler expression. We note that $t_k \neq 0$ for $k = 0, 1, \ldots, n - 2$. Indeed, if $t_k = 0$, then $x_{k+1} = x_k$ and $\nabla f(x_{k+1}) = \nabla f(x_k)$, contradicting the linear independence of the gradients. Then, $x_{k+1} = x_k + t_k d_k$ yields

$$\begin{aligned} A d_k = t_k^{-1} A (x_{k+1} - x_k) &= t_k^{-1} ((Ax_{k+1} - b) - (Ax_k - b)) \\ &= t_k^{-1} (\nabla f(x_{k+1}) - \nabla f(x_k)). \end{aligned} \tag{4.13}$$

It follows from (4.12) that, for any $k \leq n - 2$,

$$\begin{aligned} \langle -\nabla f(x_{k+1}), d_i \rangle_A &= -\nabla f(x_{k+1})^T (A d_i) = -t_i^{-1} \nabla f(x_{k+1})^T (\nabla f(x_{i+1}) - \nabla f(x_i)) \\ &= \begin{cases} 0 & \text{if } i \leq k - 1, \\ -t_k^{-1} \nabla f(x_{k+1})^T \nabla f(x_{k+1}) & \text{if } i = k. \end{cases} \end{aligned}$$

This relation, along with the expression (4.9) of t_k, transforms (4.11) into

$$d_{k+1} = -\nabla f(x_{k+1}) + \beta_{k+1} d_k, \quad k = 0, 1, \ldots, n - 2, \tag{4.14}$$

where

$$\beta_{k+1} = \frac{\nabla f(x_{k+1})^T \nabla f(x_{k+1})}{-\nabla f(x_k)^T d_k} = \frac{\nabla f(x_{k+1})^T \nabla f(x_{k+1})}{\nabla f(x_k)^T \nabla f(x_k)}. \tag{4.15}$$

The last equality follows from (4.12) and (4.14).

In summary, the *linear conjugate gradient method* is described as Algorithm 4.1.

Algorithm 4.1 Linear conjugate gradient method for solving linear system $Ax = b$ with symmetric positive definite matrix A.

Input: Symmetric positive definite matrix $A \in \mathbb{R}^{n \times n}$, vector $b \in \mathbb{R}^n$, and initial point $x_0 \in \mathbb{R}^n$.
Output: Sequence $\{x_k\} \subset \mathbb{R}^n$.
1: Compute $d_0 = -\nabla f(x_0)$, where f is given in (4.2) and $\nabla f(x) = Ax - b$.
2: **for** $k = 0, 1, 2, \ldots$ **do**
3:     Compute a step length t_k by (4.9).
4:     Compute the next point as $x_{k+1} = x_k + t_k d_k$.
5:     Compute β_{k+1} by (4.15).
6:     Compute a search direction as $d_{k+1} = -\nabla f(x_{k+1}) + \beta_{k+1} d_k$.
7: **end for**

Furthermore, the property $\nabla f(x_n) = 0$ with the assumption that $\nabla f(x_k) \neq 0$ for $k = 0, 1, \ldots, n - 1$ gives the following theorem with no assumption on the gradients.

Theorem 4.1 *Algorithm 4.1 solves the linear system $Ax = b$ with a symmetric positive definite matrix $A \in \mathbb{R}^{n \times n}$ and vector $b \in \mathbb{R}^n$ in at most n iterations.*[2]

Here, the description of Algorithm 4.1 is intended for generalizing the discussion in subsequent sections. When using the algorithm to solve a linear system, there are mathematically equivalent but more efficient ways. Although it may seem that two types of matrix-vector products Ad_k in (4.9) and Ax_{k+1} in $\nabla f(x_{k+1}) = Ax_{k+1} - b$ are required, the relation $\nabla f(x_{k+1}) = \nabla f(x_k) + t_k Ad_k$ according to (4.13) can be used to recursively compute the gradients. Therefore, we can reuse the result of Ad_k in (4.9) when computing $\nabla f(x_{k+1})$, without computing Ax_{k+1}. Other studies have developed more efficient algorithms by preconditioning (Benzi 2002; Olshanskii and Tyrtyshnikov 2014; Saad 2003); we omit these details in this book.

In addition to the linear CG method discussed above, various Krylov subspace methods have been proposed and studied, including the conjugate residual (CR) method (Stiefel 1955), which can be applied to cases in which $A \in \mathrm{Sym}(n)$ is not necessarily positive definite, and the minimum residual (MINRES) method (Paige and Saunders 1975), which is mathematically equivalent to the CR method. Furthermore, Krylov subspace methods have also been developed for cases wherein A is not symmetric, such as the generalized conjugate residual (GCR) (Eisenstat et al. 1983), generalized minimal residual (GMRES) (Saad and Schultz 1986), and biconjugate gradient (BiCG) methods (Fletcher 1976).

Remark 4.2 Here, we briefly verify that the linear CG method is indeed a Krylov subspace method, i.e., $x_k - x_0 = \sum_{l=0}^{k-1} t_l d_l$ belongs to the Krylov subspace $\mathcal{K}_k(A, r_0)$ for $k = 1, 2, \ldots, n$, where $r_0 := b - Ax_0$. We also define $r_k := b - Ax_k = -\nabla f(x_k)$.

We can, as a supplement, show $r_k, d_k \in \mathcal{K}_{k+1}(A, r_0)$ for $k = 0, 1, \ldots, n - 1$ by induction. Because $r_0 = -\nabla f(x_0) = d_0$, it is clear that $r_0, d_0 \in \mathcal{K}_1(A, r_0)$.

---

[2]In practical computation, Algorithm 4.1 may not exactly solve $Ax = b$ within n iterations due to rounding errors.

Assuming $r_k, d_k \in \mathcal{K}_{k+1}(A, r_0)$ for some $k \leq n - 2$, (4.13) gives $r_{k+1} = r_k - t_k A d_k \in \mathcal{K}_{k+2}(A, r_0)$. It then follows from (4.14) that $d_{k+1} = r_{k+1} + \beta_{k+1} d_k \in \mathcal{K}_{k+2}(A, r_0)$. Thus, we have proved $r_k, d_k \in \mathcal{K}_{k+1}(A, r_0)$ for $k = 0, 1, \ldots, n - 1$. Hence, we have $\sum_{l=0}^{k-1} t_l d_l \in \mathcal{K}_k(A, r_0)$, and the linear CG method is proved to be a Krylov subspace method.

We conclude this section by preparing for the nonlinear CG methods with a discussion of β_k. In the linear CG method, (4.15) implies several expressions for β_k. In fact, in Algorithm 4.1, it follows from (4.12) that

$$\nabla f(x_{k+1})^T \nabla f(x_{k+1}) = \nabla f(x_{k+1})^T (\nabla f(x_{k+1}) - \nabla f(x_k))$$

and

$$\nabla f(x_k)^T \nabla f(x_k) = -\nabla f(x_k)^T d_k = (\nabla f(x_{k+1}) - \nabla f(x_k))^T d_k$$

hold true. Therefore, letting $g_k := \nabla f(x_k)$ and $y_k := g_k - g_{k-1}$, we obtain two expressions of the numerator and three of the denominator for a total of six equivalent expressions of β_{k+1} as follows:

$$\begin{aligned}
\beta_{k+1} &= \frac{g_{k+1}^T g_{k+1}}{g_k^T g_k} = \frac{g_{k+1}^T g_{k+1}}{d_k^T y_{k+1}} = \frac{g_{k+1}^T g_{k+1}}{-g_k^T d_k} \\
&= \frac{g_{k+1}^T y_{k+1}}{g_k^T g_k} = \frac{g_{k+1}^T y_{k+1}}{d_k^T y_{k+1}} = \frac{g_{k+1}^T y_{k+1}}{-g_k^T d_k}.
\end{aligned} \tag{4.16}$$

These formulas lead to different types of nonlinear CG methods in the next section.

4.2 (Nonlinear) Conjugate Gradient Methods in Euclidean Spaces

We extend the linear CG method to the *nonlinear conjugate gradient methods* (or simply called *conjugate gradient methods*) for Problem 2.1 to minimize a general smooth objective function f. Specifically, we use the same formula as (4.14), i.e.,

$$d_{k+1} = -\nabla f(x_{k+1}) + \beta_{k+1} d_k, \quad k = 0, 1, \ldots \tag{4.17}$$

with $d_0 = -\nabla f(x_0)$ for computing search directions.

Step lengths (4.9) by exact line search used in the linear CG method are not necessarily available for general f. Therefore, we use approximate step lengths satisfying, e.g., the Wolfe and strong Wolfe conditions, as discussed in Sect. 2.9.1.

Regarding β_k, all expressions in (4.16) may be different values for general f. Many studies have focused on the properties of CG methods with each parameter. Some of them are listed as follows, with the uppercase letters indicating the initials of the proposing authors (except for β_k^{CD}, where CD denotes "Conjugate Descent"):

$$
\beta_k^{\mathrm{FR}} = \frac{g_k^T g_k}{g_{k-1}^T g_{k-1}}, \qquad \beta_k^{\mathrm{DY}} = \frac{g_k^T g_k}{d_{k-1}^T y_k}, \qquad \beta_k^{\mathrm{CD}} = \frac{g_k^T g_k}{-g_{k-1}^T d_{k-1}},
$$

$$
\beta_k^{\mathrm{PRP}} = \frac{g_k^T y_k}{g_{k-1}^T g_{k-1}}, \qquad \beta_k^{\mathrm{HS}} = \frac{g_k^T y_k}{d_{k-1}^T y_k}, \qquad \beta_k^{\mathrm{LS}} = \frac{g_k^T y_k}{-g_{k-1}^T d_{k-1}}, \qquad (4.18)
$$

where we put $g_k := \nabla f(x_k)$ and $y_k := g_k - g_{k-1}$, as in the previous section. They were proposed by Fletcher and Reeves (1964), Dai and Yuan (1999), Fletcher (2000), Polak and Ribière (1969) and Polyak (1969), Hestenes and Stiefel (1952), and Liu and Storey (1991), respectively. These methods were also reviewed in detail by Hager and Zhang (2006).

A more general β_k that includes the previous ones as special cases is also proposed. For example, as a generalization of the six types of β_k above,

$$
\beta_k = \frac{(1 - \lambda_k) g_k^T g_k + \lambda_k g_k^T y_k}{(1 - \mu_k - \omega_k) g_{k-1}^T g_{k-1} + \mu_k d_{k-1}^T y_k - \omega_k d_{k-1}^T g_{k-1}},
$$

called a three-parameter family, was proposed by Dai and Yuan (2001), where $\lambda_k \in [0, 1]$, $\mu_k \in [0, 1]$, and $\omega_k \in [0, 1 - \mu_k]$. For example, if $\lambda_k \equiv \mu_k \equiv \omega_k \equiv 0$, then $\beta_k = \beta_k^{\mathrm{FR}}$. Similarly, by fixing the three parameters to 0 or 1, independently of k, we can obtain the six types of β_k. Hager and Zhang (2005) proposed another β_k:

$$
\beta_k^{\mathrm{HZ}} = \frac{1}{d_{k-1}^T y_k} \left(y_k - 2 d_{k-1} \frac{\|y_k\|^2}{d_{k-1}^T y_k} \right)^T g_k.
$$

Some assumptions are required for guaranteeing the convergence properties of CG methods. For example, roughly speaking, the Fletcher–Reeves-type and Dai–Yuan-type CG methods globally converge under the strong Wolfe conditions (2.23) and (2.27) and Wolfe conditions (2.23) and (2.25), respectively. In the remainder of this chapter, we observe how these results can be extended to Riemannian manifolds.

4.3 Conjugate Gradient Methods on Riemannian Manifolds

We now discuss CG methods on a Riemannian manifold M, which we call the *Riemannian conjugate gradient methods*. The discussion in Chap. 3 suggests that we compute x_{k+1} from x_k through a retraction R on M as (3.33) and take step lengths satisfying the conditions in Sect. 3.5 to achieve generalization for Riemannian manifolds. What remains is how to compute $\beta_k \in \mathbb{R}$ and search directions.

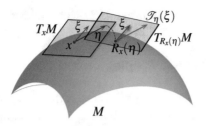

Fig. 4.1 Vector transport $\mathscr{T}$ on embedded submanifold M of $\mathbb{R}^3$. For $\eta, \xi \in T_x M$ at $x \in M$, if we simply change the starting point of ξ to $R_x(\eta)$ in $\mathbb{R}^3$ while maintaining its direction, it does not generally belong to $T_{R_x(\eta)} M$. In contrast, we have $\mathscr{T}_\eta(\xi) \in T_{R_x(\eta)} M$

In this section, we temporarily assume that the current point $x_k \in M$, search direction $\eta_k \in T_{x_k} M$, and scalar $\beta_{k+1} \in \mathbb{R}$ are given and discuss how to generalize (4.17). The first term in (4.17) can be naturally replaced with $-\operatorname{grad} f(x_{k+1}) \in T_{x_{k+1}} M$, whereas the second term seems to correspond to $\beta_{k+1} \eta_k \in T_{x_k} M$. However, in general, we cannot compute their addition since they belong to different tangent spaces. Hence, we need to map a tangent vector in $T_{x_k} M$ to one in $T_{x_{k+1}} M$. This can be realized by *parallel translation*, which is a well-known concept in differential geometry (Lee 1997; Sakai 1996), although we do not discuss it in detail in this book. However, parallel translation is sometimes difficult to compute. For example, no explicit formula is known for parallel translation on the Stiefel manifold. A wider concept called a vector transport, which includes parallel translation as a special case, together with its application to Riemannian CG methods, was proposed in Absil et al. (2008).

Using the *Whitney sum* $TM \oplus TM := \{(\eta, \xi) \mid \eta, \xi \in T_x M, \ x \in M\}$, we can define a vector transport as follows.

Definition 4.1 A map $\mathscr{T} : TM \oplus TM \to TM : (\eta, \xi) \mapsto \mathscr{T}_\eta(\xi)$ is called a *vector transport* on M if there exists a retraction R on M and $\mathscr{T}$ satisfies the following conditions for any $x \in M$:

1. $\mathscr{T}_\eta(\xi) \in T_{R_x(\eta)} M$, $\quad \eta, \xi \in T_x M$.
2. $\mathscr{T}_{0_x}(\xi) = \xi$, $\quad \xi \in T_x M$.
3. $\mathscr{T}_\eta(a\xi + b\zeta) = a\mathscr{T}_\eta(\xi) + b\mathscr{T}_\eta(\zeta)$, $\quad a, b \in \mathbb{R}, \ \eta, \xi, \zeta \in T_x M$.

The first and third conditions in Definition 4.1 mean that $\mathscr{T}_\eta$ with $\eta \in T_x M$ is a linear map from $T_x M$ to $T_{R_x(\eta)} M$, where R is said to be associated with $\mathscr{T}$. The second one implies $\mathscr{T}_{0_x} = \operatorname{id}_{T_x M}$, where 0_x is the zero vector in $T_x M$. Figure 4.1 is a conceptual illustration of a vector transport.

From Definition 4.1, if $x_{k+1} = R_{x_k}(t_k \eta_k)$, then we have $\mathscr{T}_{t_k \eta_k}(\eta_k) \in T_{x_{k+1}} M$. Thus, we can use a vector transport to map $\eta_k \in T_{x_k} M$ to a vector in $T_{x_{k+1}} M$. In other words, the formula (4.17) in the Euclidean case can be generalized to

$$\eta_{k+1} = -\operatorname{grad} f(x_{k+1}) + \beta_{k+1} \mathscr{T}_{t_k \eta_k}(\eta_k), \quad k = 0, 1, \dots . \tag{4.19}$$

Example 4.1 On a Riemannian manifold M with a retraction R, we can construct a vector transport $\mathcal{T}^R : TM \oplus TM \to TM : (\eta, \xi) \mapsto \mathcal{T}_\eta^R(\xi)$ defined by

$$\mathcal{T}_\eta^R(\xi) := DR_x(\eta)[\xi], \quad \eta, \xi \in T_x M, \ x \in M, \tag{4.20}$$

called the *differentiated retraction* (Absil et al. 2008). To verify that $\mathcal{T}^R$ is indeed a vector transport, note that $DR_x(\eta)$ is a linear map from $T_\eta T_x M \simeq T_x M$ to $T_{R_x(\eta)} M$; this proves the first and third conditions in Definition 4.1. The second condition in Definition 4.1 is equivalent to the second condition in Definition 3.16 for a retraction.

For example, we consider the Stiefel manifold $\mathrm{St}(p, n)$ endowed with the QR-based retraction R defined by (3.23). For this R, the differentiated retraction in (4.20) is written out using Proposition 3.4 as

$$\begin{aligned}
\mathcal{T}_\eta^R(\xi) = DR_X(\eta)[\xi] &= D\,\mathrm{qf}(X + \eta)[\xi] \\
&= X_+ \rho_{\mathrm{skew}}(X_+^T \xi R_+^{-1}) + (I_n - X_+ X_+^T)\xi R_+^{-1},
\end{aligned} \tag{4.21}$$

where $X \in \mathrm{St}(p, n)$, $\eta, \xi \in T_X \mathrm{St}(p, n)$, and $X + \eta = X_+ R_+$ is the QR decomposition of $X + \eta$, i.e., $X_+ := R_X(\eta) \in \mathrm{St}(p, n)$ and $R_+ := X_+^T(X + \eta)$.

When we use the formula (4.19) to compute search directions in the Riemannian CG methods, the following condition is important (Ring and Wirth 2012):

$$\left\| \mathcal{T}_{t_k \eta_k}(\eta_k) \right\|_{x_{k+1}} \leq \|\eta_k\|_{x_k}, \tag{4.22}$$

where $\eta_k \in T_{x_k} M$ and $t_k > 0$ are the search direction at the current point $x_k \in M$ and step length, respectively, and $x_{k+1} = R_{x_k}(t_k \eta_k)$. The condition (4.22) indicates that the vector transport never increases the norm of a tangent vector when the direction of the transport trajectory and the tangent vector to be transported are the same.

However, this assumption is not always valid. For example, consider the sphere S^{n-1} with the Riemannian metric (3.11) and the orthographic retraction R defined in (3.22), i.e.,

$$R_x(\eta) = \sqrt{1 - \eta^T \eta}\, x + \eta, \quad \|\eta\|_x < 1$$

for $x \in M$ and $\eta \in T_x M$. Then, the differentiated retraction $\mathcal{T}^R$ is computed as

$$\mathcal{T}_\eta^R(\xi) = DR_x(\eta)[\xi] = -\frac{\eta^T \xi}{\sqrt{1 - \eta^T \eta}} x + \xi$$

for $\xi \in T_x S^{n-1}$, whose squared norm is evaluated by $x^T x = 1$ and $x^T \xi = 0$ as

$$\|\mathcal{T}_\eta^R(\xi)\|_{R_x(\eta)}^2 = \frac{(\eta^T \xi)^2}{1 - \eta^T \eta} + \|\xi\|_x^2 \geq \|\xi\|_x^2.$$

If $\eta^T \xi \neq 0$, then $\|\mathscr{T}_\eta^R(\xi)\|_{R_x(\eta)} > \|\xi\|_x$ holds true. In particular, when $\xi_k = t_k \eta_k \neq 0$, we have $\eta_k^T \xi_k = t_k \|\eta_k\|_{x_k}^2 \neq 0$, implying that (4.22) with $\mathscr{T} = \mathscr{T}^R$ is violated. The differentiated retraction (4.21) of the QR-based retraction on the Stiefel manifold $\mathrm{St}(p, n)$ also sometimes violates the condition (4.22) unless $p = 1$.

Considering this, Sato and Iwai (2015) proposed the notion of a *scaled vector transport* $\mathscr{T}^0$ associated with a vector transport $\mathscr{T}$, which is defined as

$$\mathscr{T}_\eta^0(\xi) := \frac{\|\xi\|_x}{\|\mathscr{T}_\eta(\xi)\|_{R_x(\eta)}} \mathscr{T}_\eta(\xi), \qquad (4.23)$$

for $x \in M$ and $\eta, \xi \in T_x M$ with $\|\mathscr{T}_\eta(\xi)\|_{R_x(\eta)} \neq 0$. Note that the scaled vector transport is not a vector transport since it is not linear; i.e., it does not satisfy the third condition in Definition 4.1. However, for our purpose of computing an appropriate search direction in the Riemannian CG methods, linearity is not necessarily required.

When we replace $\mathscr{T}$ in (4.22) with $\mathscr{T}^0$, the condition (4.22) is evidently satisfied because $\|\mathscr{T}_\eta^0(\xi)\|_{R_x(\eta)} = \|\xi\|_x$. Therefore, an alternative approach in the Riemannian CG methods is to use (4.19) by replacing $\mathscr{T}$ with $\mathscr{T}^0$. However, there may exist a more appropriate way of scaling or even another approach instead of using a vector transport. For example, the way of scaling may vary depending on k in the procedure of generating $\{x_k\}$. Hence, as a general framework, we can use some maps $\mathscr{T}^{(k)} : T_{x_k} M \to T_{x_{k+1}} M$ for $k = 0, 1, \ldots$ instead of $\mathscr{T}$ in (4.19), i.e.,

$$\eta_{k+1} = -\operatorname{grad} f(x_{k+1}) + \beta_{k+1} \mathscr{T}^{(k)}(\eta_k), \quad k = 0, 1, \ldots. \qquad (4.24)$$

In summary, we propose a framework for the Riemannian CG methods as Algorithm 4.2. When we propose a new map $\mathscr{T}^{(k)}$, we should investigate carefully whether the proposed $\mathscr{T}^{(k)}$ is appropriate and analyze Algorithm 4.2 with $\mathscr{T}^{(k)}$.

Algorithm 4.2 Riemannian conjugate gradient method

Input: Riemannian manifold M, retraction R, objective function f on M, and initial point $x_0 \in M$.
Output: Sequence $\{x_k\} \subset M$.
1: Set $\eta_0 = -\operatorname{grad} f(x_0)$.
2: **for** $k = 0, 1, 2, \ldots$ **do**
3:     Compute a step length $t_k > 0$.
4:     Compute the next point as $x_{k+1} = R_{x_k}(t_k \eta_k)$.
5:     Compute $\beta_{k+1} \in \mathbb{R}$.
6:     Compute $\eta_{k+1} = -\operatorname{grad} f(x_{k+1}) + \beta_{k+1} \mathscr{T}^{(k)}(\eta_k)$ with a map $\mathscr{T}^{(k)} : T_{x_k} M \to T_{x_{k+1}} M$.
7: **end for**

The computation of $\beta_k \in \mathbb{R}$ in Algorithm 4.2 will be discussed in the subsequent sections (see also Sakai and Iiduka (2020) for a recent study). Here, we show some examples of $\mathscr{T}^{(k)}$ satisfying the desired inequality

$$\|\mathscr{T}^{(k)}(\eta_k)\|_{x_{k+1}} \leq \|\eta_k\|_{x_k}. \qquad (4.25)$$

This inequality is essential in the global convergence analysis of Riemannian CG methods. In the following examples, we assume that $x_k \in M$, $\eta_k \in T_{x_k}M$, $t_k > 0$, and $x_{k+1} = R_{x_k}(t_k\eta_k)$ are already given, and focus on defining $\mathscr{T}^{(k)}(\eta_k)$.

Example 4.2 A theoretically clear example is the use of an isometric vector transport $\mathscr{T}$, which satisfies $\|\mathscr{T}_\eta(\xi)\|_{R_x(\eta)} = \|\xi\|_x$ for any $x \in M$ and $\eta, \xi \in T_x M$. If such $\mathscr{T}$ is available, we define $\mathscr{T}^{(k)}(\eta_k) := \mathscr{T}_{t_k\eta_k}(\eta_k)$ and (4.25) is guaranteed.

Parallel translation is a theoretically natural selection that is an isometric vector transport. However, as mentioned previously, it is not easy to compute parallel translation in general. Huang et al. (2015, 2017) discussed the construction of an isometric vector transport by focusing on orthonormal bases of tangent spaces of the manifold in question. Zhu (2017) proposed a specific isometric vector transport on the Stiefel manifold $St(p, n)$ and discussed the Riemannian CG methods on $St(p, n)$ with the isometric vector transport and a nonmonotone condition for step lengths.

Example 4.3 For a retraction R and the differentiated retraction $\mathscr{T}^R$, Sato and Iwai (2015) proposed the following $\mathscr{T}^{(k)} \colon T_{x_k}M \to T_{x_{k+1}}M$:

$$
\mathscr{T}^{(k)}(\eta_k) := \begin{cases} \mathscr{T}^R_{t_k\eta_k}(\eta_k) & \text{if } \left\| \mathscr{T}^R_{t_k\eta_k}(\eta_k) \right\|_{x_{k+1}} \leq \|\eta_k\|_{x_k}, \\ \mathscr{T}^0_{t_k\eta_k}(\eta_k) & \text{otherwise,} \end{cases}
\tag{4.26}
$$

where $\mathscr{T}^0$ is the scaled vector transport (4.23) associated with $\mathscr{T}^R$. In (4.26), the differentiated retraction $\mathscr{T}^R$ is used as a vector transport when (4.22) is satisfied; the transported vector is further scaled if (4.22) is not satisfied.

Here, we note that (4.26) is well defined; i.e., when scaling, the denominator $\left\| \mathscr{T}^R_{t_k\eta_k}(\eta_k) \right\|_{x_{k+1}}$ of the scaling factor $\|\eta_k\|_{x_k} / \left\| \mathscr{T}^R_{t_k\eta_k}(\eta_k) \right\|_{x_{k+1}}$ is not 0. Indeed, we use $\mathscr{T}^0_{t_k\eta_k}(\eta_k)$ in (4.26) only when $\left\| \mathscr{T}^R_{t_k\eta_k}(\eta_k) \right\|_{x_{k+1}} > \|\eta_k\|_{x_k} \geq 0$. In addition, we can write (4.26) as $\mathscr{T}^{(k)}(\eta_k) = s_k \mathscr{T}^R_{t_k\eta_k}(\eta_k)$ with $s_k := \min \left\{ 1, \|\eta_k\|_{x_k} / \left\| \mathscr{T}^R_{t_k\eta_k}(\eta_k) \right\|_{x_{k+1}} \right\}$ to easily observe that (4.25) holds true.

The $\mathscr{T}^{(k)}$ in (4.26) is constructed based on the differentiated retraction $\mathscr{T}^R$ because it theoretically fits the (strong) Wolfe conditions well, which contain the quantity $DR_{x_k}(t_k\eta_k)[\eta_k]$, i.e., $\mathscr{T}^R_{t_k\eta_k}(\eta_k)$.

Example 4.4 Zhu and Sato (2020) recently proposed a vector transport-free approach that uses the inverse of a retraction R^{bw}, which is not necessarily the same as R used in computing $x_{k+1} = R_{x_k}(t_k\eta_k)$.

Our aim is to bring the information of $\eta_k \in T_{x_k}M$, which satisfies $x_{k+1} = R_{x_k}(t_k\eta_k)$, to $T_{x_{k+1}}M$. In the Euclidean setting with the natural retraction $R_x(\eta) = x + \eta$ in $\mathbb{R}^n$, we have $x_{k+1} = x_k + t_k\eta_k$, which implies $x_k = x_{k+1} + (-t_k\eta_k) = R_{x_{k+1}}(-t_k\eta_k)$. Therefore, in the Euclidean case, we have $\eta_k = -t_k^{-1} R_{x_{k+1}}^{-1}(x_k)$ with the canonical identification $T_{x_k}\mathbb{R}^n \simeq \mathbb{R}^n \simeq T_{x_{k+1}}\mathbb{R}^n$.

Then, again in the Riemannian case, the above discussion on the Euclidean case gives us an insight that $-t_k^{-1} R_{x_{k+1}}^{-1}(x_k) \in T_{x_{k+1}}M$ has some information of $\eta_k \in T_{x_k}M$. Furthermore, especially if x_{k+1} is sufficiently close to x_k, since a retraction is a first-order approximation of the exponential map, any retraction R^{bw} having the inverse is considered to play a role in this discussion, where "bw" stands for "backward." In other words, we consider $-t_k^{-1}\big(R_{x_{k+1}}^{\mathrm{bw}}\big)^{-1}(x_k) \in T_{x_{k+1}}M$ as a quantity related to η_k. In this case, the original retraction R is not required to have the inverse. Together with the scaling technique in Example 4.3, we define

$$\mathscr{T}^{(k)}(\eta_k) := -s_k t_k^{-1}\big(R_{R_{x_k}(t_k\eta_k)}^{\mathrm{bw}}\big)^{-1}(x_k) = -s_k t_k^{-1}\big(R_{x_{k+1}}^{\mathrm{bw}}\big)^{-1}(x_k),$$

where $s_k := \min\big\{1, \|\eta_k\|_{x_k} \big/ \big\|t_k^{-1}\big(R_{x_{k+1}}^{\mathrm{bw}}\big)^{-1}(x_k)\big\|_{x_{k+1}}\big\}$.

Earlier works on the Riemannian CG methods include Edelman et al. (1998), Edelman and Smith (1996), Smith (1994).

4.4 Fletcher–Reeves-Type Riemannian Conjugate Gradient Method

Here, we discuss the Riemannian version of the Fletcher–Reeves-type CG method together with global convergence analysis based on Ring and Wirth (2012), Sato and Iwai (2015).

4.4.1 Algorithm

Let us recall (4.18) in the Euclidean CG method; $\beta_k^{\mathrm{FR}} = (g_k^T g_k)/(g_{k-1}^T g_{k-1})$ with $g_k := \nabla f(x_k)$. This can be easily generalized to a Riemannian manifold $(M, \langle\cdot, \cdot\rangle)$ as

$$\beta_k^{\text{R-FR}} = \frac{\langle \operatorname{grad} f(x_k), \operatorname{grad} f(x_k)\rangle_{x_k}}{\langle \operatorname{grad} f(x_{k-1}), \operatorname{grad} f(x_{k-1})\rangle_{x_{k-1}}} = \frac{\|\operatorname{grad} f(x_k)\|_{x_k}^2}{\|\operatorname{grad} f(x_{k-1})\|_{x_{k-1}}^2}, \qquad (4.27)$$

where the "R-" before "FR" stands for "Riemannian." Specifically, we replace the standard inner product with the inner product defined by the Riemannian metric at each point, x_{k-1} and x_k, and the Euclidean gradients with Riemannian ones.

In this section, we focus on the Fletcher–Reeves-type Riemannian CG method, i.e., Algorithm 4.2 with (4.27). Further, we suppose a situation similar to, but more general than, that in Example 4.3. Specifically, we assume that a scaled differentiated retraction is used as $\mathscr{T}^{(k)}$ and that the inequality (4.25) for $\mathscr{T}^{(k)}$ holds true, but we do not assume the specific form (4.26). We assume, for each $k \geq 0$, that

$$\mathscr{T}^{(k)}(\eta_k) := s_k \mathscr{T}_{t_k\eta_k}^R(\eta_k) \text{ with } s_k \in (0, 1] \text{ such that } \big\|\mathscr{T}^{(k)}(\eta_k)\big\|_{x_{k+1}} \leq \|\eta_k\|_{x_k} \text{ holds.} \tag{4.28}$$

We note that the choice $s_k = \min\left\{1, \|\eta_k\|_{x_k}/\left\|\mathcal{T}^R_{t_k\eta_k}(\eta_k)\right\|_{x_{k+1}}\right\}$, i.e., (4.26) in Example 4.3, satisfies (4.28). We also assume that step lengths are chosen to satisfy the strong Wolfe conditions (3.29) and (3.31) with $0 < c_1 < c_2 < 1/2$.

4.4.2 Global Convergence Analysis

To prove the global convergence of the Fletcher–Reeves-type Riemannian CG method, we prove the following lemma. See Al-Baali (1985) for its Euclidean version.

Lemma 4.1 *In Algorithm 4.2 with $\beta_k = \beta_k^{\text{R-FR}}$ in (4.27) and (4.28), assume that t_k satisfies the strong Wolfe conditions (3.29) and (3.31) with $0 < c_1 < c_2 < 1/2$, for each $k \geq 0$. If $\operatorname{grad} f(x_k) \neq 0$ for all $k \geq 0$, then η_k is a descent direction and satisfies*

$$-\frac{1}{1-c_2} \leq \frac{\langle\operatorname{grad} f(x_k), \eta_k\rangle_{x_k}}{\|\operatorname{grad} f(x_k)\|^2_{x_k}} \leq -\frac{1-2c_2}{1-c_2}. \tag{4.29}$$

Proof In this proof, we use the notation $g_k := \operatorname{grad} f(x_k)$ to simplify the description. We prove the claim by induction. For $k = 0$, $\eta_0 = -g_0$ is clearly a descent direction and (4.29) holds because we have

$$\frac{\langle g_0, \eta_0\rangle_{x_0}}{\|g_0\|^2_{x_0}} = \frac{\langle g_0, -g_0\rangle_{x_0}}{\|g_0\|^2_{x_0}} = -1.$$

Supposing that η_k is a descent direction satisfying (4.29) for some k, we need to show that η_{k+1} is also a descent direction and satisfies (4.29) in which k is replaced with $k + 1$. Using the definition (4.27) of β_{k+1}, we have

$$\frac{\langle g_{k+1}, \eta_{k+1}\rangle_{x_{k+1}}}{\|g_{k+1}\|^2_{x_{k+1}}} = \frac{\langle g_{k+1}, -g_{k+1} + \beta_{k+1}\mathcal{T}^{(k)}(\eta_k)\rangle_{x_{k+1}}}{\|g_{k+1}\|^2_{x_{k+1}}}$$

$$= -1 + \frac{\langle g_{k+1}, \mathcal{T}^{(k)}(\eta_k)\rangle_{x_{k+1}}}{\|g_k\|^2_{x_k}}. \tag{4.30}$$

The assumption (4.28) and the strong Wolfe condition (3.31) give

$$\left|\langle g_{k+1}, \mathcal{T}^{(k)}(\eta_k)\rangle_{x_{k+1}}\right| \leq \left|\langle g_{k+1}, \mathcal{T}^R_{t_k\eta_k}(\eta_k)\rangle_{x_{k+1}}\right| \leq c_2|\langle g_k, \eta_k\rangle_{x_k}| = -c_2\langle g_k, \eta_k\rangle_{x_k}, \tag{4.31}$$

where we used the induction hypothesis that η_k is a descent direction. It follows from (4.30) and (4.31) that

$$-1 + c_2\frac{\langle g_k, \eta_k\rangle_{x_k}}{\|g_k\|^2_{x_k}} \leq \frac{\langle g_{k+1}, \eta_{k+1}\rangle_{x_{k+1}}}{\|g_{k+1}\|^2_{x_{k+1}}} \leq -1 - c_2\frac{\langle g_k, \eta_k\rangle_{x_k}}{\|g_k\|^2_{x_k}}.$$

From the induction hypothesis (4.29), i.e., $\langle g_k, \eta_k \rangle_{x_k} / \|g_k\|_{x_k}^2 \geq -(1 - c_2)^{-1}$, and the assumption $c_2 > 0$, we obtain the desired inequality

$$-\frac{1}{1 - c_2} \leq \frac{\langle g_{k+1}, \eta_{k+1} \rangle_{x_{k+1}}}{\|g_{k+1}\|_{x_{k+1}}^2} \leq -\frac{1 - 2c_2}{1 - c_2},$$

which also implies $\langle g_{k+1}, \eta_{k+1} \rangle_{x_{k+1}} < 0$. This completes the proof. $\square$

We now proceed to the global convergence property of the Fletcher–Reeves-type Riemannian CG method. The proof below is based on the Euclidean version given in Al-Baali (1985).

Theorem 4.2 *In Algorithm 4.2 with $\beta_k = \beta_k^{\text{R-FR}}$ in (4.27) and (4.28), assume that t_k satisfies the strong Wolfe conditions (3.29) and (3.31) with $0 < c_1 < c_2 < 1/2$, for each $k \geq 0$. Suppose that f is bounded below and is of class C^1. If there exists a constant $L > 0$ such that (3.35) holds, then we have*

$$\liminf_{k \to \infty} \|\operatorname{grad} f(x_k)\|_{x_k} = 0. \tag{4.32}$$

Proof We again write $g_k := \operatorname{grad} f(x_k)$. If $g_{k_0} = 0$ holds for some k_0, then we have $\beta_{k_0} = 0$ and $\eta_{k_0} = 0$ from (4.27) and (4.24); this implies $x_{k_0+1} = R_{x_{k_0}}(t_{k_0} \eta_{k_0}) = R_{x_{k_0}}(0) = x_{k_0}$. We thus obtain $g_k = 0$ for all $k \geq k_0$ and, hence, Eq. (4.32).

We now consider the case in which $g_k \neq 0$ for all $k \geq 0$. Let θ_k be the angle between $-g_k$ and η_k, i.e., $\cos \theta_k$ is defined by (3.34). It follows from (3.34) and (4.29) that

$$\cos \theta_k \geq \frac{1 - 2c_2}{1 - c_2} \frac{\|g_k\|_{x_k}}{\|\eta_k\|_{x_k}}. \tag{4.33}$$

Because the search directions are descent directions from Lemma 4.1, Zoutendijk's theorem (Theorem 3.6) guarantees (3.36), which together with (4.33) gives

$$\sum_{k=0}^{\infty} \frac{\|g_k\|_{x_k}^4}{\|\eta_k\|_{x_k}^2} < \infty. \tag{4.34}$$

Inequalities (4.29) and (4.31) are combined to give

$$\left| \langle g_k, \mathcal{T}^{(k-1)}(\eta_{k-1}) \rangle_{x_k} \right| \leq -c_2 \langle g_{k-1}, \eta_{k-1} \rangle_{x_{k-1}} \leq \frac{c_2}{1 - c_2} \|g_{k-1}\|_{x_{k-1}}^2. \tag{4.35}$$

Using (4.27), (4.28), and (4.35), we obtain the recurrence inequality for $\|\eta_k\|_{x_k}^2$:

$$\|\eta_k\|_{x_k}^2 = \left\| - g_k + \beta_k \mathscr{T}^{(k-1)}(\eta_{k-1}) \right\|_{x_k}^2$$

$$\leq \|g_k\|_{x_k}^2 + 2\beta_k \left| \langle g_k, \mathscr{T}^{(k-1)}(\eta_{k-1}) \rangle_{x_k} \right| + \beta_k^2 \left\| \mathscr{T}^{(k-1)}(\eta_{k-1}) \right\|_{x_k}^2$$

$$\leq \|g_k\|_{x_k}^2 + \frac{2c_2}{1-c_2} \beta_k \|g_{k-1}\|_{x_{k-1}}^2 + \beta_k^2 \|\eta_{k-1}\|_{x_{k-1}}^2$$

$$= \|g_k\|_{x_k}^2 + \frac{2c_2}{1-c_2} \|g_k\|_{x_k}^2 + \beta_k^2 \|\eta_{k-1}\|_{x_{k-1}}^2$$

$$= c\|g_k\|_{x_k}^2 + \beta_k^2 \|\eta_{k-1}\|_{x_{k-1}}^2, \tag{4.36}$$

where $c := (1 + c_2)/(1 - c_2) > 1$. We can successively use (4.36) with (4.27) as

$$\|\eta_k\|_{x_k}^2$$
$$\leq c\left(\|g_k\|_{x_k}^2 + \beta_k^2 \|g_{k-1}\|_{x_{k-1}}^2 + \cdots + \beta_k^2 \beta_{k-1}^2 \cdots \beta_2^2 \|g_1\|_{x_1}^2 \right) + \beta_k^2 \beta_{k-1}^2 \cdots \beta_1^2 \|\eta_0\|_{x_0}^2$$
$$= c\|g_k\|_{x_k}^4 \left(\|g_k\|_{x_k}^{-2} + \|g_{k-1}\|_{x_{k-1}}^{-2} + \cdots + \|g_1\|_{x_1}^{-2} \right) + \|g_k\|_{x_k}^4 \|g_0\|_{x_0}^{-2}$$
$$< c\|g_k\|_{x_k}^4 \sum_{j=0}^{k} \|g_j\|_{x_j}^{-2}. \tag{4.37}$$

We prove (4.32) by contradiction. To this end, assume that (4.32) does not hold. Since we are also assuming $g_k \neq 0$ for all $k \geq 0$, there exists a constant $C > 0$ such that $\|g_k\|_{x_k} \geq C > 0$ for all $k \geq 0$. Then, we have $\sum_{j=0}^{k} \|g_j\|_{x_j}^{-2} \leq C^{-2}(k+1)$. Therefore, owing to (4.37), the left-hand side of (4.34) is evaluated as

$$\sum_{k=0}^{\infty} \frac{\|g_k\|_{x_k}^4}{\|\eta_k\|_{x_k}^2} \geq \sum_{k=0}^{\infty} \frac{C^2}{c} \frac{1}{k+1} = \infty.$$

This contradicts (4.34), thereby completing the proof. □

4.5 Dai–Yuan-Type Riemannian Conjugate Gradient Method

In this section, we derive the algorithm for the Dai–Yuan-type Riemannian CG method and prove its global convergence property based on Sato (2016).

4.5.1 Algorithm

Recall that in the Euclidean case, the Dai–Yuan β_k is defined as in (4.18) by

$$\beta_k^{\mathrm{DY}} = \frac{g_k^T g_k}{d_{k-1}^T y_k},\qquad(4.38)$$

where $g_k := \nabla f(x_k)$ and $y_k := g_k - g_{k-1}$. The Dai–Yuan-type Euclidean CG method globally converges if the step lengths satisfy the Wolfe conditions; unlike in the Fletcher–Reeves-type method, the strong Wolfe conditions are not required. Here, we aim to generalize this formula to a Riemannian one, expecting a Riemannian analogy of the Dai–Yuan-type CG method to have a global convergence property if the step lengths satisfy the Wolfe conditions instead of the strong Wolfe conditions.

We again consider Algorithm 4.2. In the previous section, we generalized the Fletcher–Reeves-type β_k by replacing $g_k^T g_k$ with $\langle \mathrm{grad}\, f(x_k), \mathrm{grad}\, f(x_k)\rangle_{x_k}$, where $\mathrm{grad}\, f$ is the Riemannian gradient of f defined on M. This idea can be applied to the numerator of the fraction in the right-hand side of (4.38). However, $d_{k-1}^T y_k$ cannot be generalized this easily for the fraction's denominator.

First, $y_k = g_k - g_{k-1}$ cannot be directly generalized to the Riemannian case since $\mathrm{grad}\, f(x_k) \in T_{x_k}M$ and $\mathrm{grad}\, f(x_{k-1}) \in T_{x_{k-1}}M$ belong to different tangent spaces; i.e., we cannot perform subtraction between them. This issue may seem to be resolved by using $\mathscr{T}^{(k-1)}$, which can transport $\mathrm{grad}\, f(x_{k-1}) \in T_{x_{k-1}}M$ to $T_{x_k}M$ as $\mathscr{T}^{(k-1)}(\mathrm{grad}\, f(x_{k-1}))$.

However, if we define $y_k = \mathrm{grad}\, f(x_k) - \mathscr{T}^{(k-1)}(\mathrm{grad}\, f(x_{k-1}))$, then we encounter a new issue in generalizing the inner product-like quantity $d_{k-1}^T y_k$, i.e., $\eta_{k-1} \in T_{x_{k-1}}M$ and $y_k \in T_{x_k}M$ belong to different tangent spaces. Several approaches to resolve this issue can be considered. One is to further apply $\mathscr{T}^{(k-1)}$ to η_{k-1} and replace $d_{k-1}^T y_k$ with $\langle \mathscr{T}^{(k-1)}(\eta_{k-1}), \mathrm{grad}\, f(x_k) - \mathscr{T}^{(k-1)}(\mathrm{grad}\, f(x_{k-1}))\rangle_{x_k}$, which equals $\langle \mathscr{T}^{(k-1)}(\eta_{k-1}), \mathrm{grad}\, f(x_k)\rangle_{x_k} - \langle \mathscr{T}^{(k-1)}(\eta_{k-1}), \mathscr{T}^{(k-1)}(\mathrm{grad}\, f(x_{k-1}))\rangle_{x_k}$. Another approach is to remove $\mathscr{T}^{(k-1)}$ for the second term in this quantity, i.e., to use the formula $\langle \mathscr{T}^{(k-1)}(\eta_{k-1}), \mathrm{grad}\, f(x_k)\rangle_{x_k} - \langle \eta_{k-1}, \mathrm{grad}\, f(x_{k-1})\rangle_{x_{k-1}}$. Furthermore, if the inverse of $\mathscr{T}^{(k-1)}$, i.e., $\left(\mathscr{T}^{(k-1)}\right)^{-1}$, is available, then we may use the formula $\langle \eta_{k-1}, \left(\mathscr{T}^{(k-1)}\right)^{-1}(\mathrm{grad}\, f(x_k)) - \mathrm{grad}\, f(x_{k-1})\rangle_{x_{k-1}}$. It seems unclear which of these approaches is a more appropriate and natural extension of (4.38).

Therefore, we consider this issue from another perspective. Note that in the Euclidean space $\mathbb{R}^n$, we have another expression for β_k^{DY}, derived as

$$\begin{aligned}
\beta_k^{\mathrm{DY}} &= \beta_k^{\mathrm{DY}} \cdot \frac{\nabla f(x_{k-1})^T d_{k-1}}{\nabla f(x_{k-1})^T d_{k-1}} = \beta_k^{\mathrm{DY}} \cdot \frac{(\nabla f(x_k) - y_k)^T d_{k-1}}{\nabla f(x_{k-1})^T d_{k-1}} \\
&= \frac{-\beta_k^{\mathrm{DY}} y_k^T d_{k-1} + \beta_k^{\mathrm{DY}} \nabla f(x_k)^T d_{k-1}}{\nabla f(x_{k-1})^T d_{k-1}} = \frac{\nabla f(x_k)^T (-\nabla f(x_k) + \beta_k^{\mathrm{DY}} d_{k-1})}{\nabla f(x_{k-1})^T d_{k-1}} \\
&= \frac{\nabla f(x_k)^T d_k}{\nabla f(x_{k-1})^T d_{k-1}},
\end{aligned}\qquad(4.39)$$

where the fourth and fifth equalities follow from (4.38) and (4.17), respectively. The expression (4.39) for β_k^{DY} is useful for analyzing the global convergence of the Dai–

Yuan-type algorithm in Dai and Yuan (1999). Therefore, it will be useful to employ a formula that is a generalization of (4.39) in the Riemannian case.

The expression in (4.39) provides a clearer approach for generalization. The quantities $\nabla f(x_k)^T d_k$ and $\nabla f(x_{k-1})^T d_{k-1}$ in (4.39) are naturally replaced with $\langle \operatorname{grad} f(x_k), \eta_k \rangle_{x_k}$ and $\langle \operatorname{grad} f(x_{k-1}), \eta_{k-1} \rangle_{x_{k-1}}$, respectively. Consequently, we propose the formula

$$\beta_k^{\text{R-DY}} = \frac{\langle \operatorname{grad} f(x_k), \eta_k \rangle_{x_k}}{\langle \operatorname{grad} f(x_{k-1}), \eta_{k-1} \rangle_{x_{k-1}}}, \tag{4.40}$$

where $\beta_k^{\text{R-DY}}$ stands for the Riemannian version of β_k^{DY}. However, $\beta_k^{\text{R-DY}}$ as (4.40), like the original Euclidean formula (4.39), cannot be used in a CG algorithm because the next search direction $\eta_k \in T_{x_k} M$ is computed by using $\beta_k^{\text{R-DY}}$. In other words, the right-hand side of (4.40) in its current form can be computed only after η_k, and thus $\beta_k^{\text{R-DY}}$, are computed. Therefore, we must derive a formula of $\beta_k^{\text{R-DY}}$ that does not contain η_k.

Throughout this subsection, we assume that all quantities that appear as the denominators of fractions are nonzero. In fact, this assumption can be removed in the proposed algorithm (see Proposition 4.2). In contrast to the derivation of (4.39), we compute (4.40) using (4.24) to obtain a formula that does not contain η_k as

$$
\begin{aligned}
\beta_k^{\text{R-DY}} &= \frac{\langle \operatorname{grad} f(x_k), \eta_k \rangle_{x_k}}{\langle \operatorname{grad} f(x_{k-1}), \eta_{k-1} \rangle_{x_{k-1}}} \\
&= \frac{\left\langle \operatorname{grad} f(x_k), -\operatorname{grad} f(x_k) + \beta_k^{\text{R-DY}} \mathscr{T}^{(k-1)}(\eta_{k-1}) \right\rangle_{x_k}}{\langle \operatorname{grad} f(x_{k-1}), \eta_{k-1} \rangle_{x_{k-1}}} \\
&= \frac{-\|\operatorname{grad} f(x_k)\|_{x_k}^2 + \beta_k^{\text{R-DY}} \left\langle \operatorname{grad} f(x_k), \mathscr{T}^{(k-1)}(\eta_{k-1}) \right\rangle_{x_k}}{\langle \operatorname{grad} f(x_{k-1}), \eta_{k-1} \rangle_{x_{k-1}}}.
\end{aligned}
$$

Therefore, we conclude that

$$\beta_k^{\text{R-DY}} = \frac{\|\operatorname{grad} f(x_k)\|_{x_k}^2}{\left\langle \operatorname{grad} f(x_k), \mathscr{T}^{(k-1)}(\eta_{k-1}) \right\rangle_{x_k} - \langle \operatorname{grad} f(x_{k-1}), \eta_{k-1} \rangle_{x_{k-1}}}, \tag{4.41}$$

whose right-hand side does not contain η_k and can be used as a formula for $\beta_k^{\text{R-DY}}$.

Thus, we propose the Dai–Yuan-type Riemannian CG method as Algorithm 4.2 with $\beta_k = \beta_k^{\text{R-DY}}$ in (4.41). We again assume that $\mathscr{T}^{(k)}$ satisfies (4.28) for each $k \geq 0$, as in the Fletcher–Reeves-type Riemannian CG method. In the Dai–Yuan-type Riemannian CG method, we assume that step lengths are chosen to satisfy the Wolfe conditions (3.29) and (3.30). Then, as discussed in Sect. 4.5.2, the denominator on the right-hand side of (4.41) is proved to be nonzero; thus, (4.41) is well defined.

Remark 4.3 The formula (4.41) may seem complicated compared with the Euclidean version (4.38). However, we can rewrite this $\beta_k^{\text{R-DY}}$ as an analogy to (4.38), with the denominator rewritten using the notation $g_k := \operatorname{grad} f(x_k)$ as

$$\left\langle g_k, \mathscr{T}^{(k-1)}(\eta_{k-1})\right\rangle_{x_k} - \left\langle g_{k-1}, \eta_{k-1}\right\rangle_{x_{k-1}}$$

$$= \left\langle \mathscr{T}^{(k-1)}(\eta_{k-1}), g_k - \frac{\langle g_{k-1}, \eta_{k-1}\rangle_{x_{k-1}}}{\left\langle \mathscr{T}^{(k-1)}(g_{k-1}), \mathscr{T}^{(k-1)}(\eta_{k-1})\right\rangle_{x_k}} \mathscr{T}^{(k-1)}(g_{k-1})\right\rangle_{x_k}$$

$$= \left\langle \mathscr{T}^{(k-1)}(\eta_{k-1}), y_k\right\rangle_{x_k},$$

where we define $y_k \in T_{x_k} M$ as

$$y_k := g_k - \frac{\langle g_{k-1}, \eta_{k-1}\rangle_{x_{k-1}}}{\left\langle \mathscr{T}^{(k-1)}(g_{k-1}), \mathscr{T}^{(k-1)}(\eta_{k-1})\right\rangle_{x_k}} \mathscr{T}^{(k-1)}(g_{k-1}).$$

We thus obtain

$$\beta_k^{\text{R-DY}} = \frac{\|\operatorname{grad} f(x_k)\|_{x_k}^2}{\left\langle \mathscr{T}^{(k-1)}(\eta_{k-1}), y_k\right\rangle_{x_k}}. \tag{4.42}$$

In the Euclidean case, where $M = \mathbb{R}^n$ with the natural map $\mathscr{T}^{(k-1)} := \text{id}$, it is easy to observe $y_k = g_k - g_{k-1}$ and that (4.42) reduces to β_k^{DY} (see (4.38)).

However, unfortunately, we cannot ensure $\left\langle \mathscr{T}^{(k)}(\operatorname{grad} f(x_k)), \mathscr{T}^{(k)}(\eta_k)\right\rangle_{x_{k+1}} \neq 0$ for all $k \geq 0$ in general. Therefore, y_k, and thus the right-hand side of (4.42), are not always guaranteed to be well defined. Hence, we define $\beta_k^{\text{R-DY}}$ as (4.41) instead of (4.42) because (4.41) has an advantage of being valid without assuming $\left\langle \mathscr{T}^{(k)}(\operatorname{grad} f(x_k)), \mathscr{T}^{(k)}(\eta_k)\right\rangle_{x_{k+1}} \neq 0$ for all $k \geq 0$.

4.5.2 Global Convergence Analysis

We prove the global convergence property of the Dai–Yuan-type Riemannian CG method by using Theorem 3.6.

In the next proposition, we observe that the search direction η_k is a descent direction, i.e., $\langle \operatorname{grad} f(x_k), \eta_k\rangle_{x_k} < 0$. Furthermore, we see that the denominator on the right-hand side of (4.41) is positive. Hence, β_{k+1} is well defined by (4.41) and $\beta_{k+1} > 0$ for $k \geq 0$.

Proposition 4.2 *In Algorithm 4.2 with $\beta_k = \beta_k^{\text{R-DY}}$ in (4.41) and (4.28), assume that t_k satisfies the Wolfe conditions (3.29) and (3.30) with $0 < c_1 < c_2 < 1$, for each $k \geq 0$. If $\operatorname{grad} f(x_k) \neq 0$ for all $k \geq 0$, then this algorithm is well defined and the following two inequalities hold:*

$$\langle \operatorname{grad} f(x_k), \eta_k\rangle_{x_k} < 0, \tag{4.43}$$

$$\langle \operatorname{grad} f(x_k), \eta_k\rangle_{x_k} < \left\langle \operatorname{grad} f(x_{k+1}), \mathscr{T}^{(k)}(\eta_k)\right\rangle_{x_{k+1}}. \tag{4.44}$$

Proof We use the notation $g_k := \operatorname{grad} f(x_k)$ in this proof. The claim of this proposition is proved by induction. We first consider the case of $k = 0$. Then, inequality (4.43) clearly holds because of $\eta_0 = -g_0 \neq 0$. Further, (4.44) with $k = 0$ is obviously valid if $\langle g_1, \mathcal{T}^{(0)}(\eta_0)\rangle_{x_1} \geq 0$ because we have $\langle g_0, \eta_0\rangle_{x_0} < 0$. Otherwise, i.e., if $\langle g_1, \mathcal{T}^{(0)}(\eta_0)\rangle_{x_1} < 0$, then $\langle g_1, \mathcal{T}_{t_0\eta_0}^R(\eta_0)\rangle_{x_1} \leq \langle g_1, \mathcal{T}^{(0)}(\eta_0)\rangle_{x_1}$ holds true from the assumption (4.28). In addition, it follows from the Wolfe condition (3.30) that $\langle g_1, \mathcal{T}_{t_0\eta_0}^R(\eta_0)\rangle_{x_1} \geq c_2\langle g_0, \eta_0\rangle_{x_0}$. Therefore, noting that $c_2 < 1$ and $\langle g_0, \eta_0\rangle_{x_0} < 0$, we have

$$\langle g_1, \mathcal{T}^{(0)}(\eta_0)\rangle_{x_1} \geq \langle g_1, \mathcal{T}_{t_0\eta_0}^R(\eta_0)\rangle_{x_1} \geq c_2\langle g_0, \eta_0\rangle_{x_0} > \langle g_0, \eta_0\rangle_{x_0}.$$

In summary, (4.44) always holds true for $k = 0$.

We now suppose that, for some arbitrary $k \geq 0$, inequalities (4.43) and (4.44) hold. Note that β_{k+1} is well defined with this induction hypothesis. We can compute the left-hand side of (4.43) where k is replaced with $k + 1$ as

$$\langle g_{k+1}, \eta_{k+1}\rangle_{x_{k+1}} = \langle g_{k+1}, -g_{k+1} + \beta_{k+1}\mathcal{T}^{(k)}(\eta_k)\rangle_{x_{k+1}}$$

$$= -\|g_{k+1}\|_{x_{k+1}}^2 + \frac{\|g_{k+1}\|_{x_{k+1}}^2}{\langle g_{k+1}, \mathcal{T}^{(k)}(\eta_k)\rangle_{x_{k+1}} - \langle g_k, \eta_k\rangle_{x_k}}\langle g_{k+1}, \mathcal{T}^{(k)}(\eta_k)\rangle_{x_{k+1}}$$

$$= \frac{\|g_{k+1}\|_{x_{k+1}}^2\langle g_k, \eta_k\rangle_{x_k}}{\langle g_{k+1}, \mathcal{T}^{(k)}(\eta_k)\rangle_{x_{k+1}} - \langle g_k, \eta_k\rangle_{x_k}} < 0,$$

where the last inequality holds from the induction hypotheses (4.43) and (4.44) for k. Furthermore, from a discussion similar to the proof of $\langle g_0, \eta_0\rangle_{x_0} < \langle g_1, \mathcal{T}^{(0)}(\eta_0)\rangle_{x_1}$ above, $\langle g_{k+1}, \eta_{k+1}\rangle_{x_{k+1}} < \langle g_{k+2}, \mathcal{T}^{(k+1)}(\eta_{k+1})\rangle_{x_{k+2}}$ is always valid. Therefore, (4.43) and (4.44) are true when k is replaced with $k + 1$; thus, the proof is completed. □

Under the assumptions in Proposition 4.2, from (4.44), i.e., the positivity of the denominator on the right-hand side of (4.41), we have $\beta_k > 0$ for all $k \geq 1$.

We now proceed to the global convergence analysis of the Dai–Yuan-type Riemannian CG method. The proof is based on Sato (2016), which is analogous to the Euclidean version in Dai and Yuan (1999). In contrast with the Euclidean setting, we must carefully consider the effects of $\mathcal{T}^{(k)}$.

Theorem 4.3 *Let $\{x_k\}$ be a sequence generated by Algorithm 4.2 with $\beta_k = \beta_k^{\text{R-DY}}$ in (4.41) and (4.28), where t_k satisfies the Wolfe conditions (3.29) and (3.30) with $0 < c_1 < c_2 < 1$, for each $k \geq 0$. Suppose that f is bounded below and is of class C^1. If there exists a constant $L > 0$ such that (3.35) holds, then we have*

$$\liminf_{k\to\infty} \|\mathrm{grad}\, f(x_k)\|_{x_k} = 0. \tag{4.45}$$

Proof We again use the notation $g_k := \mathrm{grad}\, f(x_k)$ in this proof. If $g_{k_0} = 0$ for some k_0, then $\beta_{k_0} = 0$, followed by $\eta_{k_0} = 0$ and $x_{k_0+1} = R_{x_{k_0}}(0) = x_{k_0}$. It is obvious that (4.45) holds in this case.

In the following, assume $g_k \neq 0$ for all $k \geq 0$. First, note that all assumptions in Zoutendijk's theorem (Theorem 3.6) are satisfied from Proposition 4.2. Therefore, we have

$$\sum_{k=0}^{\infty} \frac{\langle g_k, \eta_k \rangle_{x_k}^2}{\|\eta_k\|_{x_k}^2} < \infty. \tag{4.46}$$

It follows from (4.24) that $\eta_{k+1} + g_{k+1} = \beta_{k+1} \mathcal{T}^{(k)}(\eta_k)$. Taking the norms, squaring, and transposing the terms, we obtain

$$\|\eta_{k+1}\|_{x_{k+1}}^2 = \beta_{k+1}^2 \|\mathcal{T}^{(k)}(\eta_k)\|_{x_{k+1}}^2 - 2\langle g_{k+1}, \eta_{k+1} \rangle_{x_{k+1}} - \|g_{k+1}\|_{x_{k+1}}^2.$$

Owing to (4.43), we can divide both sides by $\langle g_{k+1}, \eta_{k+1} \rangle_{x_{k+1}}^2$. Further, using (4.28) and (4.40), we obtain

$$\begin{aligned}
\frac{\|\eta_{k+1}\|_{x_{k+1}}^2}{\langle g_{k+1}, \eta_{k+1} \rangle_{x_{k+1}}^2} &= \frac{\|\mathcal{T}^{(k)}(\eta_k)\|_{x_{k+1}}^2}{\langle g_k, \eta_k \rangle_{x_k}^2} - \frac{2}{\langle g_{k+1}, \eta_{k+1} \rangle_{x_{k+1}}} - \frac{\|g_{k+1}\|_{x_{k+1}}^2}{\langle g_{k+1}, \eta_{k+1} \rangle_{x_{k+1}}^2} \\
&= \frac{\|\mathcal{T}^{(k)}(\eta_k)\|_{x_{k+1}}^2}{\langle g_k, \eta_k \rangle_{x_k}^2} - \left(\frac{1}{\|g_{k+1}\|_{x_{k+1}}} + \frac{\|g_{k+1}\|_{x_{k+1}}}{\langle g_{k+1}, \eta_{k+1} \rangle_{x_{k+1}}} \right)^2 + \frac{1}{\|g_{k+1}\|_{x_{k+1}}^2} \\
&\leq \frac{\|\eta_k\|_{x_k}^2}{\langle g_k, \eta_k \rangle_{x_k}^2} + \frac{1}{\|g_{k+1}\|_{x_{k+1}}^2}. \tag{4.47}
\end{aligned}$$

We prove (4.45) by contradiction. Assume that (4.45) does not hold. Since we are also assuming that $g_k \neq 0$ holds for all $k \geq 0$, there exists a constant $C > 0$ such that $\|g_k\|_{x_k} \geq C$ for all $k \geq 0$. By using the relationship (4.47) recursively, we obtain

$$\frac{\|\eta_k\|_{x_k}^2}{\langle g_k, \eta_k \rangle_{x_k}^2} \leq \sum_{i=1}^{k} \frac{1}{\|g_i\|_{x_i}^2} + \frac{\|\eta_0\|_{x_0}^2}{\langle g_0, \eta_0 \rangle_{x_0}^2} = \sum_{i=0}^{k} \frac{1}{\|g_i\|_{x_i}^2} \leq \frac{k+1}{C^2},$$

which yields

$$\sum_{k=0}^{N} \frac{\langle g_k, \eta_k \rangle_{x_k}^2}{\|\eta_k\|_{x_k}^2} \geq C^2 \sum_{k=0}^{N} \frac{1}{k+1} \to \infty, \quad N \to \infty.$$

This contradicts (4.46); thus, the proof is complete. $\square$

4.6   Numerical Examples

In this section, we show some numerical examples of the Riemannian CG methods together with the Riemannian steepest descent (SD) method by using MATLAB®.

We deal with the following optimization problem, where the Stiefel manifold $\mathrm{St}(p, n)$ with $p \leq n$ is equipped with the Riemannian metric (3.12).

Problem 4.1

$$\text{minimize} \quad f(X) := \text{tr}(X^T A X N)$$
$$\text{subject to} \quad X \in \text{St}(p, n).$$

Here, $A := \text{diag}(1, 2, \ldots, n)$ and $N := \text{diag}(p, p - 1, \ldots, 1)$.

Problem 4.1 is a special case (where A and N are specified) of Problem 5.2 and its meaning and applications are discussed in detail in Sect. 5.1.1. One of the optimal solutions to Problem 4.1 is $X_* = (I_p, 0)^T \in \text{St}(p, n)$ and the optimal value of f is $f(X_*) = \sum_{i=1}^{p} i(p - i + 1) = p(p + 1)(p + 2)/6$. When $p = 1$, Problem 4.1 reduces to the minimization problem of $f(x) = x^T A x$ on the sphere $\text{St}(1, n) = S^{n-1}$.

Throughout the section, in any optimization algorithm, we use the initial iterate $x_0 := (1, 1, \ldots, 1)^T / \sqrt{n} \in S^{n-1} = \text{St}(1, n)$ when $p = 1$, and a randomly constructed initial iterate $X_0 \in \text{St}(p, n)$ when $p > 1$ (we deal with the case of $p = 10$ later). We use the QR-based retraction (3.23) in the algorithms, which reduces to R^2 in (3.21) when $p = 1$, and the differentiated retraction $\mathcal{T}^R$ in (4.21). In the following experiments, considering the first-order necessary optimality condition (Theorem 3.4), we show the graphs of the gradient norm $\|\text{grad} f(X_k)\|_{X_k}$ for each member X_k of the generated sequences $\{X_k\} \subset \text{St}(p, n)$. We note that the function value also converged to the optimal value when the gradient norm converged to 0 in our experiments.

In Sect. 1.1.2, we considered an unconstrained optimization problem on S^2 and applied the Riemannian SD method with a fixed step length. Here, we also apply the Riemannian SD method to Problem 4.1 with $p = 1$ and $n = 3$. Note that $A = \text{diag}(1, 2, 3)$ here is different from the matrix (1.1) used in Sect. 1.1, but the situations are very similar. Because this problem is simple, in addition to a fixed step length, we can explicitly determine the step length t_k by exact search for any n. Indeed, for $x_k \in S^{n-1}$ and $\eta_k := -\text{grad} f(x_k) \neq 0$, we have

$$\phi_k(t) := f(R_{x_k}(t\eta_k)) = \frac{(x_k + t\eta_k)^T A (x_k + t\eta_k)}{\|x_k + t\eta_k\|_2^2} = \frac{a_k t^2 + b_k t + c_k}{d_k t^2 + 1},$$

where $a_k := \eta_k^T A \eta_k$, $b_k := 2x_k^T A \eta_k$, $c_k := x_k^T A x_k$, and $d_k := \eta_k^T \eta_k$. Noting that $b_k = \phi_k'(0) < 0$ and $d_k > 0$, a straightforward calculation yields

$$t_k^* := \arg\min_{t>0} \phi_k(t) = \frac{-(a_k - c_k d_k) + \sqrt{(a_k - c_k d_k)^2 + b_k^2 d_k}}{-b_k d_k}.$$

Figure 4.2a shows the results obtained by using a fixed step length $t_k \equiv 0.1$, t_k with the Armijo, Wolfe, and strong Wolfe conditions, and the step length t_k^* by exact search.

We can observe that the Armijo, Wolfe, and strong Wolfe conditions succeed to improve the performance of the Riemannian SD method. A similar experiment

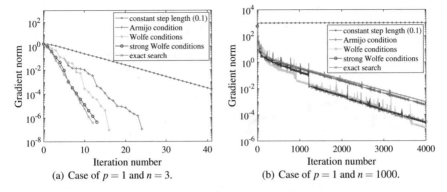

Fig. 4.2 Riemannian steepest descent methods with several step length options for Problem 4.1. The labels in the legends describe how the step lengths are chosen. In **a**, the algorithms are terminated when the gradient norm is less than 10^{-6}. In **b**, markers are put on the graphs at every 100 iterations for visibility

was also performed for a larger problem, where $p = 1$ and $n = 1000$. In this case, $\dim S^{n-1} = n - 1 = 999$. The result shown in Fig. 4.2b implies that the Riemannian SD method with an inappropriate constant step length may not solve a problem at all and that the step lengths by exact search are not necessarily the best options even if they are available. We can also observe that even when step lengths are chosen to satisfy the Armijo, Wolfe, or strong Wolfe conditions, the Riemannian SD method suffers from slow convergence as n increases.

Next, we compare the Riemannian SD with Riemannian CG methods. We again solve Problem 4.1 with $p = 1$ and $n = 1000$ and apply the Riemannian SD and CG methods. The results shown in Fig. 4.3 demonstrate that the CG methods generally converge significantly faster than SD. However, CG (FR) with the Wolfe conditions, which is not guaranteed to converge, exhibits a slower convergence than CG (FR) with the strong Wolfe conditions and CG (DY) with the (strong) Wolfe conditions.

We focus on the Riemannian CG methods. As a larger problem, we deal with Problem 4.1 with $p = 10$ and $n = 300$, where $\dim \mathrm{St}(p, n) = np - p(p + 1)/2 = 2945$ is larger than the dimension of the manifolds in the previous examples. In Sect. 4.5.1, we discussed several possibilities of generalizing the denominator of the Euclidean Dai–Yuan formula (4.38). Among them, an alternative that was not adopted leads to $\beta_k' = \|\mathrm{grad}\, f(x_k)\|_{x_k}^2 / \langle \mathscr{T}^{(k-1)}(\eta_{k-1}), \mathrm{grad}\, f(x_k) - \mathscr{T}^{(k-1)}(\mathrm{grad}\, f(x_{k-1}))\rangle_{x_k}$. The results in Fig. 4.4 show that the use of β_k' instead of $\beta_k^{\mathrm{R\text{-}DY}}$ in (4.41) does not perform well in addition to CG (FR) with the Wolfe conditions. In contrast, we observe again that CG (FR) with the strong Wolfe conditions and CG (DY) with the (strong) Wolfe conditions, whose global convergence properties are guaranteed, perform well.

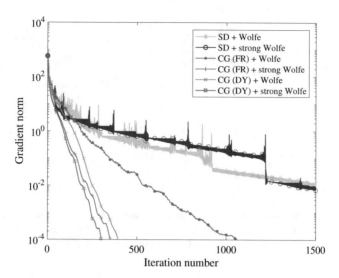

Fig. 4.3 Case of $p = 1$ and $n = 1000$. In the legend, SD refers to Algorithm 3.3; CG (FR) refers to Algorithm 4.2 with (4.26) and $\beta_k = \beta_k^{\text{R-FR}}$ in (4.27); and CG (DY) refers to Algorithm 4.2 with (4.26) and $\beta_k = \beta_k^{\text{R-DY}}$ in (4.41). Further, "+ Wolfe" and "+ strong Wolfe" mean that step lengths are chosen to satisfy the Wolfe and strong Wolfe conditions, respectively. Markers are put on the graphs at every 50 iterations

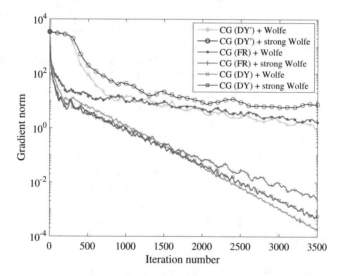

Fig. 4.4 Case of $p = 10$ and $n = 300$. In addition to the notation in Fig. 4.3, CG (DY') is Algorithm 4.2 with (4.26) and $\beta_k = \beta_k'$. Markers are put on the graphs at every 100 iterations

Chapter 5
Applications of Riemannian Optimization

In this chapter, we introduce some practical examples of Riemannian optimization. The introduced problems are themselves important; furthermore, they can provide readers with insights into other Riemannian optimization problems and help them in solving such problems that they may encounter. Several libraries have been developed for implementing Riemannian optimization, including Manopt in MATLAB® (Boumal et al. 2014), Pymanopt in Python (Townsend et al. 2016), ManifoldOptim in R (Martin et al. 2016), ROPTLIB in C++ (Huang et al. 2018b), and Manopt.jl in Julia (Bergmann 2019). By using such libraries, we can easily utilize Riemannian optimization algorithms.

5.1 Numerical Linear Algebra

In numerical linear algebra, eigenvalue and singular value problems are of great importance owing to their rich applications. In this section, we discuss the formulations of the decomposition problems as Riemannian optimization problems, some properties of these problems, and optimization algorithms for solving them.

5.1.1 Eigenvalue Decomposition

We recall the example introduced in Sect. 1.1.2, which is an optimization problem on the two-dimensional sphere S^2. Generalizing it to that on the $(n-1)$-dimensional sphere $S^{n-1} := \{x \in \mathbb{R}^n \mid x^T x = 1\}$, we consider the following optimization problem.

© Springer Nature Switzerland AG 2021

H. Sato, *Riemannian Optimization and Its Applications*,
SpringerBriefs in Control, Automation and Robotics,
https://doi.org/10.1007/978-3-030-62391-3_5

Problem 5.1

$$\text{minimize} \quad f(x) := x^T A x$$

$$\text{subject to} \quad x \in S^{n-1}.$$

Here, A is an $n \times n$ symmetric matrix that is not necessarily positive definite.

In fact, this problem is equivalent to finding a nonzero $x \in \mathbb{R}^n$ that minimizes the function $g \colon \mathbb{R}^n - \{0\} \to \mathbb{R}$ defined by

$$g(x) := \frac{x^T A x}{x^T x}$$

because x and αx with any nonzero $\alpha \in \mathbb{R}$ attain the same function value, i.e.,

$$g(\alpha x) = g(x) \tag{5.1}$$

holds true. Thus, essentially, only the direction of x is significant. Therefore, we can restrict the Euclidean norm of x to 1, i.e., $x \in S^{n-1}$. Here, the function $g(x)$ is called the *Rayleigh quotient* of A. Because the Euclidean gradient of g can be computed as

$$\nabla g(x) = \frac{2}{(x^T x)^2}((x^T x) A x - (x^T A x) x) = \frac{2}{x^T x}(A x - g(x) x),$$

$\nabla g(x) = 0$ holds if and only if $A x = g(x) x$. Therefore, solving Problem 5.1 is equivalent to computing an eigenvector associated with the smallest eigenvalue of A.

Remark 5.1 One may wonder why we consider Problem 5.1 instead of the problem of minimizing the Rayleigh quotient in the whole $\mathbb{R}^n - \{0\}$. A possible answer to this question is the failure of Euclidean Newton's method for minimizing $g(x)$. By computing the Euclidean gradient and Hessian matrix of g, it can be shown that the Newton equation $\nabla^2 g(x)[d] = -\nabla g(x)$ with respect to $d \in \mathbb{R}^n$, where x is fixed, has a unique solution $d = x$ unless $g(x)$ is an eigenvalue of A (Absil et al. 2008). Therefore, the search direction in Newton's method at x_k is $d_k = x_k$. When the step length is fixed to 1, we obtain $x_{k+1} = x_k + d_k = x_k + x_k = 2x_k$, which implies $x_k = 2^k x_0$ with initial point $x_0 \neq 0$. Thus, x_k is generated on the ray $\{t x_0 \mid t > 0\}$, and from (5.1) we have $g(x_k) = g(x_0)$ for all $k \geq 0$, i.e., the function value never changes in the procedure of Newton's method. Therefore, Newton's method for this problem fails unless x_0 is an optimal solution, which is clearly meaningless because such a case would mean that we know the solution beforehand.

Why does Newton's method fail to obtain an optimal solution? In this case, the failure cannot be attributed to the local convergence property of Newton's method. Indeed, the above discussion concludes that it always fails, however close to an optimal solution we take x_0, unless x_0 is exactly an optimal solution. In fact, an assumption for Newton's method, i.e., the positive definiteness of the Hessian matrix $\nabla^2 g(x_*)$ at an optimal solution x_*, is violated in this problem. This fact can be shown without explicitly computing $\nabla^2 g(x_*)$ as follows: Since (5.1) holds true for any $\alpha \neq 0$ and an arbitrarily fixed x, differentiating both sides with respect to α yields $x^T \nabla g(\alpha x) = 0$. Further differentiating both sides with respect to α, we obtain $x^T \nabla^2 g(\alpha x) x = 0$. Since $\alpha \neq 0$ and $x \neq 0$ are arbitrary, setting $\alpha = 1$ and $x = x_*$, we get the relation $x_*^T \nabla^2 g(x_*) x_* = 0$. Because $x_* \neq 0$, this indicates that $\nabla^2 g(x_*)$ is not positive definite. See also Absil et al. (2008) for a discussion related to this problem.

Focusing on the fact that $x \in S^{n-1}$ is regarded as an $n \times 1$ matrix, we can extend Problem 5.1 to a problem with respect to $X = (x_1, x_2, \ldots, x_p) \in \mathbb{R}^{n \times p}$ with $1 \leq p \leq n$ and $x_i \in \mathbb{R}^n$ being the ith column vector of X. Imposing that the column vectors are mutually orthonormal, we have the condition $X^T X = I_p$, i.e., X is on the Stiefel manifold $\mathrm{St}(p, n)$. Thus, we obtain the following optimization problem.

Problem 5.2

$$\text{minimize} \quad F(X) := \mathrm{tr}(X^T A X N)$$
$$\text{subject to} \quad X \in \mathrm{St}(p, n).$$

Here, $A \in \mathrm{Sym}(n)$ and $N := \mathrm{diag}(\mu_1, \mu_2, \ldots, \mu_p)$ with $\mu_1 > \mu_2 > \cdots > \mu_p > 0$.

The diagonal matrix N plays a role in appropriately ordering the columns of X. To see this, we first observe the following proposition (Absil et al. 2008; Brockett 1993).

Proposition 5.1 *In Problem 5.2, $X = (x_1, x_2, \ldots, x_p) \in \mathrm{St}(p, n)$ is a critical point of F if and only if each column $x_i \in \mathbb{R}^n$ is an eigenvector of A.*

We prove this proposition after deriving the Riemannian gradient of F. Proposition 5.1 yields the following corollary, which shows that the diagonal matrix N orders the columns of an optimal solution $X_* \in \mathrm{St}(p, n)$.

Corollary 5.1 *In Problem 5.2, assume that $A \in \mathrm{Sym}(n)$ has eigenvalues $\lambda_1, \lambda_2, \ldots, \lambda_n$ with $\lambda_1 \leq \lambda_2 \leq \cdots \leq \lambda_n$. Then, $X = (x_1, x_2, \ldots, x_p) \in \mathrm{St}(p, n)$ is an optimal solution to the problem if and only if the ith column x_i is an eigenvector of A associated with λ_i for $i = 1, 2, \ldots, p$.*

Proof We first note that X is an optimal solution to the problem if and only if $\mathrm{grad}\, F(X) = 0$ holds true and X attains the smallest value of $F(X)$ among the critical points of F. It follows from Proposition 5.1 that the condition $\mathrm{grad}\, F(X) = 0$ is equivalent to $AX = X \,\mathrm{diag}(d_1, d_2, \ldots, d_p)$ for some eigenvalues $d_1, d_2, \ldots, d_p \in \mathbb{R}$ of A. Denoting $D = \mathrm{diag}(d_1, d_2, \ldots, d_p)$, we have

$$F(X) = \mathrm{tr}(X^T X D N) = \mathrm{tr}(DN) = \sum_{i=1}^{p} d_i \mu_i. \tag{5.2}$$

Since $X \in \mathrm{St}(p, n)$ implies that $x_1, x_2, \ldots, x_p$ are mutually orthonormal, $d_1, d_2, \ldots, d_p$ are obtained by permuting p values from $\lambda_1, \lambda_2, \ldots, \lambda_n$. Considering

$\mu_1 > \mu_2 > \cdots > \mu_p > 0$, we conclude that (5.2) for a critical point X attains its minimum if and only if $d_i = \lambda_i$ for $i = 1, 2, \ldots, p$. This is the desired conclusion.
□

Here, we briefly describe the requisites to implement the Riemannian steepest descent and conjugate gradient methods for Problem 5.2. We require a retraction, which is independent of the objective function f, and the Riemannian gradient of f, which depends on f and the Riemannian metric on $\mathrm{St}(p, n)$.

As discussed before, we can use the QR-based retraction R in (3.23), the differentiated retraction $\mathscr{T}^R$ in (4.21) to define $\mathscr{T}^{(k)}$ in (4.26), and the Riemannian metric (3.12). Furthermore, let $\bar{F}: \mathbb{R}^{n \times p} \to \mathbb{R}$ be defined as $\bar{F}(X) := \mathrm{tr}(X^T A X N)$, whose restriction to $\mathrm{St}(p, n)$ is F. Now, the directional derivative with respect to arbitrary $V \in \mathbb{R}^{n \times p}$ is computed as

$$\frac{d}{dt} \bar{F}(X + tV)\Big|_{t=0} = \mathrm{tr}(V^T A X N) + \mathrm{tr}(X^T A V N) = \mathrm{tr}((2AXN)^T V),$$

noting that A is symmetric. We thus obtain $\nabla \bar{F}(X) = 2AXN$. See Remark 2.3 and Example 2.15 for a similar discussion on deriving the Euclidean gradient of a function defined on a matrix space. Then, from Proposition 3.3 and (3.18), the Riemannian gradient $\mathrm{grad}\, F$ can be calculated via the orthogonal projection onto the tangent space as, for any $X \in \mathrm{St}(p, n)$,

$$\mathrm{grad}\, F(X) = P_X(\nabla \bar{F}(X)) = 2(AXN - X \,\mathrm{sym}(X^T AXN)). \tag{5.3}$$

We can prove Proposition 5.1 by using the Riemannian gradient in (5.3).

Proof (*Proposition* 5.1) We assume that X is a critical point of F. Then, we have

$$AXN = X \,\mathrm{sym}(X^T AXN). \tag{5.4}$$

Multiplying X^T from the left gives $X^T AXN = \mathrm{sym}(X^T AXN)$. Hence, $X^T AXN$ is symmetric, implying $X^T AXN = NX^T AX$. For the (i, j) element $(X^T AX)_{ij}$ of $X^T AX$, we have $(X^T AX)_{ij} \mu_j = \mu_i (X^T AX)_{ij}$. If $i \neq j$, then $\mu_i \neq \mu_j$ holds true from the assumption of the proposition, yielding $(X^T AX)_{ij} = 0$. Hence, $X^T AX$ is a diagonal matrix. Furthermore, it follows from (5.4) that $AXN = XX^T AXN$, i.e., $AX = XX^T AX$ since N is invertible. Therefore, for each column x_i of X, we have $Ax_i = (X^T AX)_{ii} x_i$. Noting that $X \in \mathrm{St}(p, n)$ implies $x_i \neq 0$, we conclude that x_i is an eigenvector of A.

Conversely, if each column of X is an eigenvector of A, then $AX = XD$ for some diagonal matrix $D \in \mathbb{R}^{n \times n}$. Since DN is diagonal (and hence, symmetric), we obtain

$$\mathrm{grad}\, F(X) = 2(XDN - X \,\mathrm{sym}(X^T XDN)) = 2(XDN - X \,\mathrm{sym}(DN)) = 0.$$

Thus, the proof is complete.   □

5.1.2 Singular Value Decomposition

In this subsection, we deal with singular value decomposition, mostly based on Sato and Iwai (2013). The singular value decomposition of $A \in \mathbb{R}^{m \times n}$ can be partly achieved by solving the following optimization problem.

Problem 5.3

$$\text{minimize} \quad f(U, V) := \text{tr}(U^T A V N)$$
$$\text{subject to} \quad (U, V) \in \text{St}(p, m) \times \text{St}(p, n).$$

Here, $1 \leq p \leq n \leq m$ and $N := \text{diag}(\mu_1, \mu_2, \ldots, \mu_p)$ with $\mu_1 > \mu_2 > \cdots > \mu_p > 0$.

Specifically, we have the following proposition, whose proof can be found in Sato and Iwai (2013).

Proposition 5.2 *Let m, n, and p be positive integers satisfying $m \geq n \geq p$; A be an $m \times n$ matrix; and (U_*, V_*) be a global optimal solution to Problem 5.3. Then, the ith columns of U_* and V_* for $i = 1, 2, \ldots, p$ are left- and right-singular vectors of A associated with the ith largest singular value, respectively.*

Remark 5.2 In Sato and Iwai (2013), the problem of maximizing f, i.e., minimizing $-\text{tr}(U^T A V N)$, is addressed. If (U, V) is on $\text{St}(p, m) \times \text{St}(p, n)$, then so are $(-U, V)$ and $(U, -V)$, and we have $f(U, V) = -f(-U, V) = -f(U, -V)$. Therefore, if (U_*, V_*) is a global minimizer of f, then $(-U_*, V_*)$ and $(U_*, -V_*)$ are global maximizers of f. Thus, minimizing and maximizing f essentially have almost no difference.

Before discussing $\text{St}(p, m) \times \text{St}(p, n)$, we briefly review a general theory of product manifolds. Given two maps $f_1 \colon X_1 \to Y_1$ and $f_2 \colon X_2 \to Y_2$, we define the map $f_1 \times f_2 \colon X_1 \times X_2 \to Y_1 \times Y_2$ as $(f_1 \times f_2)(x_1, x_2) := (f_1(x_1), f_2(x_2))$ for any $(x_1, x_2) \in X_1 \times X_2$. Let M_1 and M_2 be smooth manifolds with atlases $\{(U_\alpha, \varphi_\alpha)\}_{\alpha \in A}$ and $\{(V_\beta, \psi_\beta)\}_{\beta \in B}$, respectively. Then, the product space $M_1 \times M_2$ with the product topology is shown to be a second-countable Hausdorff space. Furthermore, $\mathcal{A} := \{(U_\alpha \times V_\beta, \varphi_\alpha \times \psi_\beta)\}_{(\alpha, \beta) \in A \times B}$ is shown to be a C^∞ atlas of $M_1 \times M_2$. Thus, $M_1 \times M_2$ with the atlas $\mathcal{A}$ is a smooth manifold, called a *product manifold* of M_1 and M_2. For the projections $\pi_1 \colon M_1 \times M_2 \to M_1$ and $\pi_2 \colon M_1 \times M_2 \to M_2$ defined by $\pi_1(x_1, x_2) := x_1$ and $\pi_2(x_1, x_2) := x_2$ for any $x_1 \in M_1$ and $x_2 \in M_2$, the map $\zeta \mapsto (\mathrm{D}\pi_1(x_1, x_2)[\zeta], \mathrm{D}\pi_2(x_1, x_2)[\zeta])$ for $\zeta \in T_{(x_1, x_2)}(M_1 \times M_2)$ is shown to be an isomorphism between $T_{(x_1, x_2)}(M_1 \times M_2)$ and $T_{x_1} M_1 \times T_{x_2} M_2$. We can thus identify $T_{(x_1, x_2)}(M_1 \times M_2)$ with $T_{x_1} M_1 \times T_{x_2} M_2$. See also Tu (2011) for details.

Let $M := \text{St}(p, m) \times \text{St}(p, n)$ be the product manifold of $\text{St}(p, m)$ and $\text{St}(p, n)$. We can endow M with a Riemannian metric based on each manifold as

$$\langle (\xi_1, \xi_2), (\eta_1, \eta_2) \rangle_{(U,V)} := \langle \xi_1, \eta_1 \rangle_U + \langle \xi_2, \eta_2 \rangle_V = \text{tr}(\xi_1^T \eta_1) + \text{tr}(\xi_2^T \eta_2)$$

for $(U, V) \in M$ and (ξ_1, ξ_2), $(\eta_1, \eta_2) \in T_{(U,V)}M \simeq T_U \operatorname{St}(p, m) \times T_V \operatorname{St}(p, n)$. Here, $\langle \xi_1, \eta_1 \rangle_U$ and $\langle \xi_2, \eta_2 \rangle_V$ are the inner products defined by the Riemannian metrics of the form (3.12) on $\operatorname{St}(p, m)$ and $\operatorname{St}(p, n)$, respectively.

As in the previous subsection, we require a retraction, the Riemannian gradient grad f on $\operatorname{St}(p, m) \times \operatorname{St}(p, n)$, and possibly a vector transport to implement the Riemannian steepest descent and conjugate gradient methods.

For retractions R^1 on $\operatorname{St}(p, m)$ and R^2 on $\operatorname{St}(p, n)$, a retraction R^M on M is given by

$$R_{(U,V)}^M(\xi_1, \xi_2) := (R_U^1(\xi_1), R_V^2(\xi_2)) \in M \tag{5.5}$$

for $(U, V) \in M$ and $(\xi_1, \xi_2) \in T_{(U,V)}M$. If we choose the QR-based retractions of the form (3.23) as R^1 and R^2, then we obtain a specific retraction R^M on M as $R_{(U,V)}^M(\xi_1, \xi_2) = (\operatorname{qf}(U + \xi_1), \operatorname{qf}(V + \xi_2))$. Based on a similar discussion, we can define a vector transport $\mathscr{T}^M$ on M with vector transports $\mathscr{T}^1$ and $\mathscr{T}^2$ on the Stiefel manifolds $\operatorname{St}(p, m)$ and $\operatorname{St}(p, n)$, respectively, as

$$\mathscr{T}_{(\eta_1, \eta_2)}^M(\xi_1, \xi_2) := \left(\mathscr{T}_{\eta_1}^1(\xi_1), \mathscr{T}_{\eta_2}^2(\xi_2) \right) \tag{5.6}$$

for $(U, V) \in M$ and (η_1, η_2), $(\xi_1, \xi_2) \in T_{(U,V)}M$. A specific option for $\mathscr{T}^M$ in (5.6) is obtained when we choose, e.g., the differentiated retractions of the form (4.21) as both $\mathscr{T}^1$ and $\mathscr{T}^2$. With the identification $T_U \operatorname{St}(p, m) \times T_V \operatorname{St}(p, n) \simeq T_{(U,V)}M$, it is straightforward to show that R^M and $\mathscr{T}^M$ defined by (5.5) and (5.6) are indeed a retraction and vector transport, respectively.

We define $\bar{f}$ as $\bar{f}(U, V) := \operatorname{tr}(U^T A V N)$ on $\mathbb{R}^{m \times p} \times \mathbb{R}^{n \times p}$, whose restriction to $\operatorname{St}(p, m) \times \operatorname{St}(p, n)$ is equal to f. The Euclidean gradient of $\bar{f}$ is written as

$$\nabla \bar{f}(U, V) = (\nabla_U \bar{f}(U, V), \nabla_V \bar{f}(U, V)) = (AVN, A^T UN),$$

where $\nabla_U \bar{f}$ is the Euclidean gradient of $\bar{f}$ with respect to U considering that V is fixed; $\nabla_V \bar{f}$ is defined in the same manner. The Riemannian gradient of f is then written using the orthogonal projections P_U and P_V to $T_U \operatorname{St}(p, m)$ and $T_V \operatorname{St}(p, n)$, respectively, as

$$\operatorname{grad} f(U, V) = (P_U(\nabla_U \bar{f}(U, V)), \ P_V(\nabla_V \bar{f}(U, V)))$$
$$= (AVN - U \operatorname{sym}(U^T AVN), \ A^T UN - V \operatorname{sym}(V^T A^T UN)). \tag{5.7}$$

This is because $P_{(U,V)} \colon \mathbb{R}^{m \times p} \times \mathbb{R}^{n \times p} \to T_U \operatorname{St}(p, m) \times T_V \operatorname{St}(p, n) \simeq T_{(U,V)}M$ defined by $P_{(U,V)}(W, Z) := (P_U(W), P_V(Z))$ is the orthogonal projection to $T_{(U,V)}M$.

We conclude this subsection with a remark on the complex singular value decomposition.

Remark 5.3 A complex matrix $A \in \mathbb{C}^{m \times n}$ with $m \geq n$ can be decomposed into $A = U \Sigma V^*$, called the complex singular value decomposition, where $U \in \mathbb{C}^{m \times n}$

and $V \in \mathbb{C}^{n \times n}$ satisfy $U^*U = V^*V = I_n$, and Σ is an $n \times n$ diagonal matrix with nonnegative diagonal elements (Golub and Van Loan 2013). Here, $\cdot^*$ denotes the Hermitian conjugation, i.e., $B^* := \bar{B}^T$ for any complex matrix B. Note that $\bar{B}$ is the complex conjugate of B. The *complex Stiefel manifold*

$$\mathrm{St}(p, n; \mathbb{C}) := \{X \in \mathbb{C}^{n \times p} \mid X^*X = I_p\}$$

plays an important role in complex singular value decomposition regarding the p largest singular values. A discussion analogous to that in this subsection can derive a complex singular value decomposition algorithm based on optimization on the product of two complex Stiefel manifolds. See Sato (2014), Sato and Iwai (2013) for more details.

5.2 Control Theory

Helmke and Moore (1994) discussed dynamical systems pertaining to gradient flows on Riemannian manifolds and related optimization, a discrete-time version of which leads to Riemannian optimization algorithms. In Helmke and Moore (1994), the applications of the methods include control theory. In this section, we introduce some examples of applications of Riemannian optimization to control theory.

In control engineering, the properties of the controlled object are formulated by equations to construct a mathematical model by focusing on the input–output relationship of the control system. A *state-space representation* is often used to mathematically model the system in question. In this section, we focus on a continuous time-invariant linear system, which is modeled as

$$\begin{cases} \dot{x}(t) = Ax(t) + Bu(t), \\ y(t) = Cx(t), \end{cases} \tag{5.8}$$

where $x(t) \in \mathbb{R}^n$, $y(t) \in \mathbb{R}^q$, and $u(t) \in \mathbb{R}^p$ are called the *state*, *output*, and *input vectors* of the system, respectively. The matrices $A \in \mathbb{R}^{n \times n}$, $B \in \mathbb{R}^{n \times p}$, and $C \in \mathbb{R}^{q \times n}$ are constants that characterize the system.[1] In what follows, we assume that $A + A^T$ is negative definite, i.e., $-(A + A^T)$ is positive definite, in (5.8), which holds true for, e.g., dissipative systems (see Ishizaki et al. (2015), Sato and Sato (2016) for details). Under this assumption, we can show that A is *stable*, i.e., the real parts of the eigenvalues of A are all negative. Then, the system is guaranteed to be *asymptotically stable*.

---

[1]The second equation in (5.8) may be replaced with $y(t) = Cx(t) + Du(t)$ with $D \in \mathbb{R}^{q \times p}$ in a more general case, and the matrices A, B, C, and D may not be constants, but they vary depending on time t in a time-variant system.

There are various topics for the model (5.8), some of which are related to Riemannian optimization. For example, system identification algorithms via Riemannian optimization can be found in Sato and Sato (2017), Sato et al. (2020), Usevich and Markovsky (2014). An important topic in facilitating the controller design involves model reduction, which reduces the dimension of the system state. In this section, we mainly focus on an optimal model reduction problem from the perspective of Riemannian optimization. Other applications in control engineering and robotics can be found in, e.g., Lee and Park (2018), Sato (2017), Traversaro et al. (2016).

5.2.1 Optimal Model Reduction

In association with the original system (5.8), we consider the reduced system

$$\begin{cases} \dot{x}_r(t) = A_r x_r(t) + B_r u(t), \\ y_r(t) = C_r x_r(t), \end{cases} \tag{5.9}$$

where $r < n$, $A_r \in \mathbb{R}^{r \times r}$, $B_r \in \mathbb{R}^{r \times p}$, and $C_r \in \mathbb{R}^{q \times r}$. Our aim is to determine A_r, B_r, and C_r to construct a reduced model that effectively approximates the original system (5.8).

To discuss what we should minimize, we give a brief introduction of some basic concepts. The *Laplace transform $F(s)$* of a function $f(t)$ on $[0, \infty)$, which is denoted as $\mathscr{L}[f(t)]$, is defined by

$$F(s) := \int_0^\infty f(t) e^{-st} dt, \quad s \in \mathbb{C}.$$

The operator $\mathscr{L}[\cdot]$ is linear, and if f is smooth, then we have

$$\mathscr{L}[f'(t)] = s\mathscr{L}[f(t)] - f(0).$$

Let $X(s) = \mathscr{L}[x(t)]$, $Y(s) = \mathscr{L}[y(t)]$, and $U(s) = \mathscr{L}[u(t)]$ be the Laplace transforms of $x(t)$, $y(t)$, and $u(t)$, respectively, where $\mathscr{L}$ acts on each vector component. Assuming that the initial values of the state, output, and input are the zero vectors, we take the Laplace transform of (5.8) to obtain

$$\begin{cases} sX(s) = AX(s) + BU(s), \\ Y(s) = CX(s). \end{cases} \tag{5.10}$$

When $sI_n - A$ is invertible, (5.10) implies that

$$Y(s) = CX(s) = C(sI_n - A)^{-1}BU(s) = G(s)U(s),$$

where

$$G(s) := C(sI_n - A)^{-1}B, \quad s \in \mathbb{C}$$

is called the *transfer function matrix* of the system. Whereas the state-space representation describes the relationships between the state, input, and output vectors, the transfer function expresses the relationship between the input and output. The H_2 *norm* of the system is then defined by its transfer function G as

$$\|G\|_{H_2} := \sqrt{\frac{1}{2\pi} \int_{-\infty}^{\infty} \mathrm{tr}(G(i\omega)^*G(i\omega))d\omega},$$

where i denotes the imaginary unit. When two systems with transfer functions G_1 and G_2 have a common input u, assuming that

$$\int_0^{\infty} \|u(t)\|_2^2 dt \le 1,$$

i.e., the energy of the input is bounded, we can prove that

$$\sup_{t \ge 0}\|y_1(t) - y_2(t)\|_2 \le \|G_1 - G_2\|_{H_2}$$

holds (Gugercin et al. 2008; Sato 2019), where y_1 and y_2 are the outputs of the two systems. Hence, it can be said that the two systems are close to each other if $\|G_1 - G_2\|_{H_2}$ is small.

Let G and G_r be the transfer functions of the original and reduced systems, respectively, i.e., $G(s) := C(sI_n - A)^{-1}B$ and $G_r(s) := C_r(sI_r - A_r)^{-1}B_r$. We aim to minimize the H_2 norm of the difference G_e of G and G_r, which is written as

$$G_e(s) := G(s) - G_r(s) = C(sI_n - A)^{-1}B - C_r(sI_r - A_r)^{-1}B_r$$
$$= C_e(sI_{n+r} - A_e)^{-1}B_e,$$

where we define $A_e := \begin{pmatrix} A & 0 \\ 0 & A_r \end{pmatrix}$, $B_e := \begin{pmatrix} B \\ B_r \end{pmatrix}$, and $C_e := (C, -C_r)$.

We have several form options that can be assumed for matrices A_r, B_r, and C_r. Here, following Sato and Sato (2015), Yan et al. (1999), we assume that the matrices of the reduced model are transformations of the original matrices by $U \in \mathbb{R}^{n \times r}$ with $U^T U = I_r$, i.e., $U \in \mathrm{St}(r, n)$, as

$$A_r = U^T A U, \qquad B_r = U^T B, \qquad C_r = CU. \tag{5.11}$$

Our objective function J of U on the Stiefel manifold $\mathrm{St}(r, n)$ is now defined by

$$J(U) := \|G_e(s)\|_{H_2}^2 = \|C(sI_n - A)^{-1}B - CU(sI_r - U^TAU)^{-1}U^TB\|_{H_2}^2.$$
$$(5.12)$$

Because we assume that $A + A^T$ is negative definite and $U^TU = I_r$, we can observe that the matrix $A_r + A_r^T = U^T(A + A^T)U$ is also negative definite. Therefore, $A_r = U^TAU$, as well as A, is stable. Accordingly, the objective function J can be written as $J(U) = \text{tr}(C_e E_c C_e^T) = \text{tr}(B_e^T E_o B_e)$, where E_c and E_o are the *controllability* and *observability Gramians*, respectively. They are the solutions to the following Lyapunov equations (Yan et al. 1999):

$$A_e E_c + E_c A_e^T + B_e B_e^T = 0, \qquad A_e^T E_o + E_o A_e + C_e^T C_e = 0. \qquad (5.13)$$

We define $E_c =: \begin{pmatrix} \Sigma_c & X \\ X^T & P \end{pmatrix}$ and $E_o =: \begin{pmatrix} \Sigma_o & Y \\ Y^T & Q \end{pmatrix}$, where $\Sigma_c, \Sigma_o \in \mathbb{R}^{n \times n}$ are the controllability and observability Gramians of the original system (5.8), respectively, $P, Q \in \text{Sym}(r)$, and $X, Y \in \mathbb{R}^{n \times r}$. We can rewrite (5.13) using these matrices as

$$A\Sigma_c + \Sigma_c A^T + BB^T = A^T \Sigma_o + \Sigma_o A + C^T C = 0,$$
$$U^TAUP + PU^TA^TU + U^TBB^TU = U^TA^TUQ + QU^TAU + U^TC^TCU = 0,$$
$$(5.14)$$

$$AX + XU^TA^TU + BB^TU = A^TY + YU^TAU - C^TCU = 0. \qquad (5.15)$$

The objective function J can then be written as

$$J(U) = \text{tr}(C^TC(\Sigma_c + UPU^T - 2XU^T)) = \text{tr}(BB^T(\Sigma_o + UQU^T + 2YU^T)).$$

Thus, we obtain the following optimization problem on the Stiefel manifold.

Problem 5.4

$$\text{minimize} \quad J(U) := \|G(s) - G_r(s)\|_{H_2}^2$$
$$\text{subject to} \quad U \in \text{St}(r, n).$$

Remark 5.4 The objective function J has $\mathcal{O}_r$ invariance, i.e.,

$$J(UW) = J(U), \quad W \in \mathcal{O}_r, \qquad (5.16)$$

where $\mathcal{O}_r$ is the orthogonal group. This can be shown by using (5.12) as

$$J(UW)$$
$$= \|C(sI_n - A)^{-1}B - C(UW)(s(W^TW) - (UW)^TA(UW))^{-1}(UW)^TB\|_{H_2}^2$$
$$= \|C(sI_n - A)^{-1}B - CU(WW^T)(sI_r - U^TAU)^{-1}(WW^T)U^TB\|_{H_2}^2$$
$$= J(U).$$

Equation (5.16) implies that when U_* is an optimal solution to Problem 5.4, so is $U_* W$ for any $W \in \mathcal{O}_r$. This means that the optimal solutions are not isolated, which can be harmful, especially if we use second-order optimization methods. Therefore, it may be more effective to identify $U W$ with U and discuss optimization on the quotient manifold $\mathrm{St}(r, n)/\mathcal{O}_r$, i.e., Grassmann manifold $\mathrm{Grass}(r, n)$ (see Sect. 3.1.5). Note that the dimension of $\mathrm{Grass}(r, n)$ is smaller than that of $\mathrm{St}(r, n)$, as $\dim \mathrm{Grass}(r, n) = r(n - r) < r(n - r) + r(r - 1)/2 = \dim \mathrm{St}(r, n)$ unless $r = 1$.

Let $[U] := \{U W \mid W \in \mathcal{O}_r\} \in \mathrm{Grass}(r, n)$ denote the equivalence class represented by $U \in \mathrm{St}(r, n)$ and $\tilde{J}$ be a function on $\mathrm{Grass}(r, n)$ defined by $\tilde{J}([U]) := J(U)$. Note that $\tilde{J}$ is well defined since J has $\mathcal{O}_r$ invariance. Then, the resultant optimization problem on $\mathrm{Grass}(r, n)$ is written as follows.

Problem 5.5

$$\text{minimize} \quad \tilde{J}([U]) := J(U)$$
$$\text{subject to} \quad [U] \in \mathrm{Grass}(r, n).$$

For further discussion on the H_2 optimal model reduction problem on the Grassmann manifold, refer to Sato and Sato (2015).

Remark 5.5 Another approach for finding A_r, B_r, and C_r for the reduced model (5.9) involves freely changing B_r and C_r without assuming specific forms while assuming that A_r has the form $A_r = U^T A U$ for some $U \in \mathrm{St}(r, n)$, as in (5.11), to guarantee the stability of A_r. The resultant problem, whose decision variables are $(U, B_r, C_r) \in \mathrm{St}(r, n) \times \mathbb{R}^{r \times p} \times \mathbb{R}^{q \times r}$, may provide a solution leading to better A_r, B_r, and C_r that Problem 5.4 cannot attain. This approach is discussed in detail in Sato and Sato (2016).

We return to Problem 5.4 and derive some requisites for Riemannian optimization. We endow $\mathrm{St}(r, n)$ with the Riemannian metric (3.12), where p is replaced with r in this case. For first-order methods, such as the steepest descent and conjugate gradient methods, we compute the Riemannian gradient of J. Defining $\bar{J} : \mathbb{R}^{n \times r} \to \mathbb{R}$ as $\bar{J}(U) = \mathrm{tr}(C^T C(\Sigma_c + U P U^T - 2 X U^T))$, whose restriction to $\mathrm{St}(r, n)$ is equal to $J : \mathrm{St}(r, n) \to \mathbb{R}$, the Euclidean gradient $\nabla \bar{J}(U)$ of $\bar{J}$ at U can be written as (Yan et al. 1999)

$$\nabla \bar{J}(U) = 2((-C^T C + A^T U Y^T)X + (C^T C U + A^T U Q)P$$
$$+ (B B^T + A U X^T)Y + (B B^T U + A U P)Q).$$

Because $U^T A U$ is stable, it follows from (5.14) and (5.15) that P, Q, X, and Y are smooth functions of U (Dullerud and Paganini 2000). Let P_ξ, Q_ξ, X_ξ, and Y_ξ denote the directional derivatives of P, Q, X, and Y at U along ξ, respectively. From (5.14) and (5.15), the derivatives P_ξ, Q_ξ, X_ξ, and Y_ξ are the solutions to the following Sylvester equations:

$$U^T A U P_\xi + P_\xi U^T A^T U + S_P = 0, \quad U^T A^T U Q_\xi + Q_\xi U^T A U + S_Q = 0,$$
$$A X_\xi + X_\xi U^T A^T U + S_X = 0, \quad A^T Y_\xi + Y_\xi U^T A U + S_Y = 0,$$

where

$$S_P = 2\,\mathrm{sym}(\xi^T A U P + U^T A \xi P + \xi^T B B^T U),$$
$$S_Q = 2\,\mathrm{sym}(\xi^T A^T U Q + U^T A^T \xi Q + \xi^T C^T C U),$$
$$S_X = X \xi^T A^T U + X U^T A^T \xi + B B^T \xi, \quad S_Y = Y \xi^T A U + Y U^T A \xi - C^T C \xi.$$

Then, the gradient grad $J(U)$ of J at $U \in \mathrm{St}(r, n)$ can be computed using the orthogonal projection onto $T_U \mathrm{St}(r, n)$ (see (3.19)) as

$$\mathrm{grad}\, J(U) = P_U(\nabla \bar{J}(U)) = (I_n - UU^T)\nabla \bar{J}(U) + U\,\mathrm{skew}(U^T \nabla \bar{J}(U)). \quad (5.17)$$

Since $\bar{J}(UW) = \bar{J}(U)$ for any $W \in \mathcal{O}_r$, fixing U, the function $K(W) := \bar{J}(UW)$ of $W \in \mathcal{O}_r$ is a constant function on $\mathcal{O}_r$. Therefore, for any curve $W(t)$ on $\mathcal{O}_r$ satisfying $W(0) = I_r \in \mathcal{O}_r$ and $\dot{W}(0) = \xi \in T_{I_r}\mathcal{O}_r$, we have

$$0 = \frac{d}{dt} K(W(t)) \Big|_{t=0} = \mathrm{D}K(W(0))[\dot{W}(0)]$$
$$= \mathrm{D}\bar{J}(U)[U\xi] = \mathrm{tr}(\nabla \bar{J}(U)^T U\xi) = \mathrm{tr}((U^T \nabla \bar{J}(U))^T \xi).$$

Noting that $T_{I_r}\mathcal{O}_r = \{\xi \in \mathbb{R}^{r \times r} \mid \xi^T + \xi = 0\} = \mathrm{Skew}(r)$, we observe that ξ is an arbitrary skew-symmetric matrix. This implies that $U^T \nabla \bar{J}(U)$ is symmetric, i.e., $\mathrm{skew}(U^T \nabla \bar{J}(U)) = 0$, which simplifies (5.17) into

$$\mathrm{grad}\, J(U) = (I_n - UU^T)(\nabla \bar{J}(U)). \quad (5.18)$$

By using a retraction, possibly a vector transport, and the Riemannian gradient (5.18), we can derive the steepest descent and conjugate gradient methods for Problem 5.4. In Sato and Sato (2015), the Riemannian Hessian of J on $\mathrm{St}(r, n)$ is further derived and the trust-region method for Problem 5.4 is proposed. Similarly, the geometry of Problem 5.5 on $\mathrm{Grass}(r, n)$ is also discussed in detail.

Some other discussions on optimal model reduction algorithms based on Riemannian optimization can be found in, e.g., Sato (2019), Sato and Sato (2018).

5.3 Statistical Methods

Riemannian optimization also has a wide variety of applications in statistics. For example, the maximum likelihood estimation for Gaussian mixture models, which is usually performed by the expectation-maximization (EM) method, is alternatively attained by batch and stochastic Riemannian optimization (Hosseini and Sra 2015, 2020a) (also refer to Sect. 6.1 for stochastic optimization).

In the following subsection, as another example, we focus on an application of Riemannian optimization to canonical correlation analysis.

5.3.1 Canonical Correlation Analysis

Let $x_1, x_2, \ldots, x_m$ and $y_1, y_2, \ldots, y_n$ be random variables, T be the sample size, and $X \in \mathbb{R}^{T \times m}$ and $Y \in \mathbb{R}^{T \times n}$ be the corresponding data matrices. For simplicity, we assume that the sample means of $x_1, x_2, \ldots, x_m$ and $y_1, y_2, \ldots, y_n$ are all 0.

We consider the linear combinations of $x_1, x_2, \ldots, x_m$ and $y_1, y_2, \ldots, y_n$ as $f := Xu$ and $g := Yv$ with coefficients $u \in \mathbb{R}^m$ and $v \in \mathbb{R}^n$, respectively. The sample correlation coefficient between f and g is written by using the variances $\mathrm{Var}(f)$ and $\mathrm{Var}(g)$ and covariance $\mathrm{Cov}(f, g)$ as

$$\rho = \frac{\mathrm{Cov}(f, g)}{\sqrt{\mathrm{Var}(f)}\sqrt{\mathrm{Var}(g)}} = \frac{u^T C_{XY} v}{\sqrt{u^T C_X u}\sqrt{v^T C_Y v}}, \tag{5.19}$$

where $C_X := X^T X$, $C_Y := Y^T Y$, and $C_{XY} := X^T Y$. *Canonical correlation analysis seeks u and v that maximize the value of ρ.* The matrices C_X and C_Y are, by definition, symmetric and positive semidefinite. Further, we assume that they are positive definite in the following.

An approach to maximizing ρ involves the use of the transformation $u' = \sqrt{C_X} u$ and $v' = \sqrt{C_Y} v$, and rewriting ρ as

$$\rho = \frac{u'^T \left(\sqrt{C_X}^{-1} C_{XY} \sqrt{C_Y}^{-1}\right) v'}{\sqrt{u'^T u'}\sqrt{v'^T v'}}.$$

Then, u'_* and v'_* that maximize ρ are obtained as left- and right-singular vectors associated with the largest singular value of the matrix $\sqrt{C_X}^{-1} C_{XY} \sqrt{C_Y}^{-1}$, which yields that $(u_*, v_*) := \left(\sqrt{C_X}^{-1} u'_*, \sqrt{C_Y}^{-1} v'_*\right)$ is a maximizer of (5.19).

As another approach without directly computing the singular value decomposition of $\sqrt{C_X}^{-1} C_{XY} \sqrt{C_Y}^{-1}$, we can consider ρ in (5.19) as a function of u and v. As (u, v) and $(\alpha u, \beta v)$ with arbitrary nonzero scalars α and β attain the same value of ρ, the norms of u and v do not affect the value of ρ. We endow $\mathbb{R}^m$ and $\mathbb{R}^n$ with inner products, as shown in Example 2.1, with the symmetric positive definite matrices C_X and C_Y, respectively. The induced norms of u and v are $\sqrt{u^T C_X u}$ and $\sqrt{v^T C_Y v}$, respectively. Then, it is natural to consider the case where the norms of u and v are 1, which is equivalent to the condition that u and v are on $S_{C_X}^{m-1}$ and $S_{C_Y}^{n-1}$, where S_G^{n-1} is defined as $S_G^{n-1} := \{x \in \mathbb{R}^n \mid x^T G x = 1\}$ for an $n \times n$ symmetric positive definite matrix G. The resultant optimization problem is written as follows.

Problem 5.6

$$\text{maximize} \quad u^T C_{XY} v$$
$$\text{subject to} \quad (u, v) \in S_{C_X}^{m-1} \times S_{C_Y}^{n-1}.$$

The maximum value of the objective function is called the first *canonical correlation coefficient*.

We recall the generalized Stiefel manifold $\text{St}_G(p, n) := \{X \in \mathbb{R}^{n \times p} \mid X^T G X = I_p\}$ (see Example 3.12). We can generalize Problem 5.6 to obtain up to the pth canonical correlation coefficient as the following problem on the product manifold of two generalized Stiefel manifolds $\text{St}_{C_X}(p, m)$ and $\text{St}_{C_Y}(p, n)$, where $p \leq \min\{m, n\}$.

Problem 5.7

$$\text{maximize} \quad f(U, V) := \text{tr}\left(U^T C_{XY} V N\right) \tag{5.20}$$
$$\text{subject to} \quad (U, V) \in \text{St}_{C_X}(p, m) \times \text{St}_{C_Y}(p, n).$$

Here, $N := \text{diag}(\sigma_1, \sigma_2, \ldots, \sigma_p)$ with $\sigma_1 > \sigma_2 > \cdots > \sigma_p > 0$.

Remark 5.6 Similar to Remark 5.2 for Problem 5.3, maximizing and minimizing f in Problem 5.7 essentially have almost no difference. To obtain a solution to Problem 5.7, we can alternatively minimize $-f$. Another approach is the minimization of f to obtain a minimizer $(\tilde{U}_*, \tilde{V}_*)$ and defining (U_*, V_*) as $(-\tilde{U}_*, \tilde{V}_*)$ or $(\tilde{U}_*, -\tilde{V}_*)$, which is then an optimal solution to Problem 5.7. In the following discussion, we derive the Riemannian gradient of f.

Although canonical correlation analysis can be achieved by algorithms from numerical linear algebra, performing it by solving the optimization problem is effective, especially when adaptively analyzing the data. Adaptive canonical correlation analysis via Riemannian optimization is discussed in Yger et al. (2012).

As in the previous sections, we require some tools, such as a retraction and the Riemannian gradient, to implement Riemannian optimization algorithms for Problem 5.7. We focus on the generalized Stiefel manifold $\text{St}_G(p, n)$ with $p \leq n$ and a general positive definite matrix $G \in \text{Sym}(n)$. Then, the product manifold $\text{St}_{C_X}(p, m) \times \text{St}_{C_Y}(p, n)$ can be easily handled, as in Sect. 5.1.2, where we discussed the product manifold $\text{St}(p, m) \times \text{St}(p, n)$ based on the results on $\text{St}(p, n)$.

Let us discuss the geometry of $\text{St}_G(p, n)$. Focusing on the positive definite matrix $G \in \text{Sym}(n)$, we endow $\mathbb{R}^{n \times p}$ with a Riemannian metric $\langle \cdot, \cdot \rangle$ defined by

$$\langle \xi, \eta \rangle_X := \text{tr}(\xi^T G \eta), \quad \xi, \eta \in T_X \mathbb{R}^{n \times p} \simeq \mathbb{R}^{n \times p}, \quad X \in \mathbb{R}^{n \times p}. \tag{5.21}$$

Then, $\text{St}_G(p, n)$ can be equipped with the induced metric from (5.21), and thus be a Riemannian submanifold of $(\mathbb{R}^{n \times p}, \langle \cdot, \cdot \rangle)$. A discussion similar to the case of the Stiefel manifold $\text{St}(p, n)$ yields that the orthogonal projection P_X onto the tangent space $T_X \text{St}_G(p, n) = \{\xi \in \mathbb{R}^{n \times p} \mid X^T G \xi + \xi^T G X = 0\}$ acts on $Y \in \mathbb{R}^{n \times p}$ as

$$P_X(Y) = Y - X \operatorname{sym}(X^T G Y).$$

Let $\bar{g}$ and $g := \bar{g}|_{\operatorname{St}_G(p,n)}$ be a smooth function on $\mathbb{R}^{n \times p}$ and its restriction to $\operatorname{St}_G(p, n)$, respectively. Then, for any $\xi \in T_X \mathbb{R}^{n \times p} \simeq \mathbb{R}^{n \times p}$, we have

$$\mathrm{D}\bar{g}(X)[\xi] = \operatorname{tr}(\nabla\bar{g}(X)^T \xi) = \operatorname{tr}((G^{-1}\nabla\bar{g}(X))^T G\xi) = \langle G^{-1}\nabla\bar{g}(X), \xi \rangle_X,$$

where $\nabla\bar{g}(X) = \left(\frac{\partial \bar{g}}{\partial X_{ij}}(X)\right) \in \mathbb{R}^{n \times p}$ is the Euclidean gradient with respect to the standard inner product. Since $G^{-1}\nabla\bar{g}(X) \in \mathbb{R}^{n \times p} \simeq T_X \mathbb{R}^{n \times p}$, we conclude that the Riemannian gradient of $\bar{g}$ at X on $\mathbb{R}^{n \times p}$ is $\operatorname{grad}\bar{g}(X) = G^{-1}\nabla\bar{g}(X)$. From Proposition 3.3, the Riemannian gradient of g at X on $\operatorname{St}_G(p, n)$ is then obtained as

$$\begin{aligned}
\operatorname{grad} g(X) &= P_X(\operatorname{grad}\bar{g}(X)) = \operatorname{grad}\bar{g}(X) - X \operatorname{sym}(X^T G \operatorname{grad}\bar{g}(X)) \\
&= G^{-1}\nabla\bar{g}(X) - X \operatorname{sym}(X^T \nabla\bar{g}(X)). \tag{5.22}
\end{aligned}$$

Based on the aforementioned discussion, we endow the product manifold $M := \operatorname{St}_{C_X}(p, m) \times \operatorname{St}_{C_Y}(p, n)$ with the Riemannian metric

$$\langle (\xi_1, \xi_2), (\eta_1, \eta_2) \rangle_{(U,V)} := \operatorname{tr}(\xi_1^T C_X \eta_1) + \operatorname{tr}(\xi_2^T C_Y \eta_2)$$

for $(U, V) \in M$ and $(\xi_1, \xi_2), (\eta_1, \eta_2) \in T_{(U,V)}M$. We define $\bar{f}: \mathbb{R}^{m \times p} \times \mathbb{R}^{n \times p} \to \mathbb{R}$ as $\bar{f}(U, V) := \operatorname{tr}(U^T C_{XY} V N)$, whose restriction to M is equal to f in (5.20). The Euclidean gradient of $\bar{f}$ is written as

$$\nabla\bar{f}(U, V) = (\nabla_U \bar{f}(U, V), \nabla_V \bar{f}(U, V)) = (C_{XY} V N, C_{XY}^T U N).$$

Therefore, similar to the discussion for deriving (5.7), the Riemannian gradient of f can be obtained from (5.22) as

$$\begin{aligned}
&\operatorname{grad} f(U, V) \\
&= (C_X^{-1} C_{XY} V N - U \operatorname{sym}(U^T C_{XY} V N), \; C_Y^{-1} C_{XY}^T U N - V \operatorname{sym}(V^T C_{XY}^T U N)).
\end{aligned}$$

When retractions R_1 and R_2 and vector transports $\mathcal{T}_1$ and $\mathcal{T}_2$ on $\operatorname{St}_{C_X}(p, m)$ and $\operatorname{St}_{C_Y}(p, n)$, respectively, are available, we can construct a retraction R and vector transport $\mathcal{T}$ on M as in (5.5) and (5.6). A retraction on the generalized Stiefel manifold based on the polar decomposition is discussed in Yger et al. (2012) along with its use in iterative optimization algorithms. As another retraction on $\operatorname{St}_G(p, n)$, we recall Example 3.12, where a retraction on $\operatorname{St}_G(p, n)$ is introduced as

$$R_X^G(\eta) = \sqrt{G}^{-1} \operatorname{qf}\left(\sqrt{G}(X + \eta)\right) \tag{5.23}$$

for $X \in \operatorname{St}_G(p, n)$ and $\eta \in T_X \operatorname{St}_G(p, n)$. A vector transport on $\operatorname{St}_G(p, n)$ is also obtained as the differentiated retraction $\mathcal{T}^{R^G}$:

$$\mathscr{T}_\eta^{R^G}(\xi) = \mathrm{D}R_X^G(\eta)[\xi] = \sqrt{G}^{-1}\mathrm{D}\,\mathrm{qf}\left(\sqrt{G}(X+\eta)\right)[\sqrt{G}\xi]$$
$$= \sqrt{G}^{-1}\mathrm{D}R_{X'}(\eta')[\xi'] = \sqrt{G}^{-1}\mathscr{T}_{\eta'}^R(\xi'),$$

where $X' := \sqrt{G}X \in \mathrm{St}(p,n)$, $\eta' := \sqrt{G}\eta$, $\xi' := \sqrt{G}\xi \in T_{X'}\mathrm{St}(p,n)$, and R and $\mathscr{T}^R$ are the retraction and vector transport on $\mathrm{St}(p,n)$ defined in (3.23) and (4.21), respectively.

Theoretically, we have all requisites for implementing the Riemannian steepest descent or conjugate gradient methods for Problem 5.7. However, the computation of (5.23) seems more complex than that of (3.23) on $\mathrm{St}(p,n)$ due to the effect of G. In the remainder of this section, we discuss an efficient implementation of R^G in (5.23) following the discussion in Sato and Aihara (2019).

Note that the right-hand side of (5.23) contains $\sqrt{G}$ and its inverse, which leads to a computational cost of $O(n^3)$. To avoid the comparatively intensive computation of $\sqrt{G}$ and $\sqrt{G}^{-1}$, Sato and Aihara (2019) successfully proposed the following algorithm using the concept of *Cholesky QR decomposition*. Cholesky QR decomposition is an approach for achieving QR decomposition using Cholesky decomposition. Assume that we attempt to seek the QR decomposition of an $n \times p$ full-rank matrix A with $p \le n$. Theoretically, let $A = QR$ be the QR decomposition of A, i.e., $Q \in \mathrm{St}(p,n)$ is what we aim to find, and R is a $p \times p$ upper triangular matrix. Then, we obtain $A^T A = (QR)^T(QR) = R^T R$ using $Q^T Q = I_p$, where $A^T A = R^T R$ is actually the Cholesky decomposition of the $p \times p$ matrix $A^T A$. Based on the uniqueness of Cholesky decomposition, we can obtain $Q = AR^{-1}$ after obtaining R by Cholesky decomposition of $A^T A$. In other words, Cholesky QR decomposition is performed by the Cholesky decomposition of $A^T A$ as $A^T A = R^T R$, followed by the inverse transformation $Q = AR^{-1}$.

Here, we apply these steps to the computation of the map qf on the right-hand side of (5.23). To this end, we find the Cholesky decomposition of

$$\left(\sqrt{G}(X+\eta)\right)^T \left(\sqrt{G}(X+\eta)\right) = (X+\eta)^T G(X+\eta)$$

as $R^T R$, where R is a $p \times p$ upper triangular matrix. Then, the Q-factor of $\sqrt{G}(X+\eta)$ can be computed as in the above discussion by qf $\left(\sqrt{G}(X+\eta)\right) = \sqrt{G}(X+\eta)R^{-1}$. Therefore, we have

$$R_X^G(\eta) = \sqrt{G}^{-1}\sqrt{G}(X+\eta)R^{-1} = (X+\eta)R^{-1}.$$

The overall algorithm for computing $R_X^G(\eta)$ is summarized in Algorithm 5.1.

Algorithm 5.1 Cholesky QR-based retraction on generalized Stiefel manifold

---

Input: Positive definite $G \in \mathrm{Sym}(n)$, point $X \in \mathrm{St}_G(p, n)$, and tangent vector $\eta \in T_X \mathrm{St}_G(p, n)$.
Output: Retracted point $R_X^G(\eta)$.
1: Compute $Z = (X + \eta)^T G(X + \eta)$.
2: Compute the Cholesky decomposition of Z as $Z = R^T R$.
3: Compute $R_X^G(\eta) = (X + \eta)R^{-1}$.

The computations in Algorithm 5.1 neither contain $\sqrt{G}$ nor its inverse. Consequently, the highest computation in the algorithm is the multiplication of $G \in \mathbb{R}^{n \times n}$ and $X + \eta \in \mathbb{R}^{n \times p}$, whose computational cost ($O(n^2 p)$) is less than $O(n^3)$ required for directly computing (5.23). In addition, it is natural to assume $p \ll n$ since p is the number of canonical correlation coefficients to be obtained.

From the example in this section, it can be observed that the selection of a retraction as well as its computation are important in practical computation.

5.4 Further Applications

In this chapter, we mainly introduced applications of Riemannian optimization on the (generalized) Stiefel and Grassmann manifolds. It should be noted that there are other manifolds as well that are used in applications. In this section, we briefly review the further applications of Riemannian optimization and introduce the manifolds on which optimization problems are formulated.

Topics of numerical linear algebra themselves have many applications in other fields. Riemannian optimization is also applied to such problems beyond eigenvalue and singular value problems. For example, algorithms for joint diagonalization (Absil et al. 2006; Sato 2017; Theis et al. 2009; Yamada and Ezaki 2003) and joint singular value decomposition (Sato 2015) are derived as optimization methods based on the Stiefel manifold, orthogonal group, or *oblique manifold* $\mathcal{OB}(p, n) := \{X \in \mathbb{R}^{n \times p} \mid (X^T X)_{ii} = 1, \ i = 1, 2, \ldots, p\}$, which have applications in, e.g., independent component analysis (Afsari and Krishnaprasad 2004) and blind source separation (Congedo et al. 2011). Some other applications of joint diagonalization and joint singular value decomposition can be found in, e.g., Hori (2010), Pfeffer et al. (2019). Furthermore, some studies apply Riemannian optimization to nonlinear or inverse eigenvalue problems (Yao et al. 2016, 2020; Zhao et al. 2015). In Yao et al. (2016), an algorithm for doubly stochastic inverse eigenvalue problems, which seek to find a doubly stochastic matrix whose eigenvalues are equal to given complex numbers $\lambda_1, \lambda_2, \ldots, \lambda_n$, is proposed based on the Fletcher–Reeves-type Riemannian conjugate gradient method, which was discussed in Sect. 4.4. The manifold discussed in Yao et al. (2016) is the product manifold $\mathcal{OB}(n, n) \times \mathcal{O}_n \times \mathcal{U}$, where $\mathcal{U} := \{U \in \mathbb{R}^{n \times n} \mid U_{ij} = 0, \ (i, j) \in \mathcal{I}\}$ and $\mathcal{I}$ is an index subset determined by the given $\lambda_1, \lambda_2, \ldots, \lambda_n$. This example implies that we can create a new manifold to manage an optimization problem even if the feasible set of the problem is not an already known manifold.

Low-rank matrix completion (Mishra et al. 2014; Vandereycken 2013) and low-rank tensor completion (Kasai and Mishra 2016; Kressner et al. 2014; Steinlechner 2016) are studied as optimization problems based on the fixed-rank manifold, i.e., the Riemannian manifold of fixed-rank matrices, and the manifold of tensors of fixed multilinear rank, respectively. Low-rank matrix completion is applied to, e.g., collaborative filtering for recommender systems and pattern recognition. A manifold structure for low-rank tensors and its various applications are reviewed in Grasedyck et al. (2013).

Nonnegative matrix factorization (NMF) is a popular approach used in fields such as computer vision, signal processing, and machine learning (Lee and Seung 1999). NMF (approximately) factorizes a given matrix V into $V \approx WH$, where all elements of V, W, and H are nonnegative. The orthogonal NMF problem is an NMF problem with an additional orthogonal constraint on a factor matrix and can be formulated as an optimization problem on the Stiefel manifold (He et al. 2020; Pompili et al. 2014).

The manifold of fixed-rank symmetric positive semidefinite matrices (Journée et al. 2010; Massart and Absil 2020) and the manifold of symmetric positive definite matrices (Bhatia 2007; Moakher 2005) are also important and have been studied. They can be applied to, e.g., the maximal cut of a graph, computer vision, and signal processing.

See also Chap. 6 where several other Riemannian optimization methods and their applications are introduced.

Chapter 6
Recent Developments in Riemannian Optimization

In this chapter, we review the recent developments in Riemannian optimization, such as stochastic and constrained optimization. A few other topics, including second-order and nonsmooth optimization, are also briefly reviewed. Interested readers may refer to the references introduced in the subsequent sections.

6.1 Stochastic Optimization

Recently, machine learning for big data has attracted increased attention as a part of the development of artificial intelligence. Machine learning is broadly divided into two categories: supervised learning and unsupervised learning. Supervised learning attempts to estimate the appropriate outputs for given inputs using training data, which are pairs of inputs and true outputs given beforehand, whereas unsupervised learning extracts information such as data structures using only inputs.

This section focuses on optimization problems with an objective function consisting of a large number of component functions, such as a function that includes information about the data appearing in supervised learning. As an example, in a regression problem, we seek a regression function to estimate the outputs for the given inputs. Another example is a classification problem, where we seek a classifier, which returns a category for the input object. For the training data, the regression function or classifier is determined such that it minimizes the training error between the estimated and true outputs. Hence, in both problems, we must solve an optimization problem. We need enough data to attain accurate learning; however, increasing the sample size enlarges the scale of the optimization problem due to the expensive computational cost for training errors. For large-scale optimization problems, it is not easy to even compute the value or gradient of the objective function at a point.

© Springer Nature Switzerland AG 2021

H. Sato, *Riemannian Optimization and Its Applications*,
SpringerBriefs in Control, Automation and Robotics,
https://doi.org/10.1007/978-3-030-62391-3_6

Stochastic optimization algorithms randomly sample a part of data at each iteration and compute an effective search direction using the sampled data.

Specifically, in this section, we focus on the problem with the objective function f of the form $f(x) := \frac{1}{N} \sum_{i=1}^{N} f_i(x)$, where N is the sample size, which is assumed to be considerably large in the following discussion. Various studies have been conducted on problems where f is defined in a Euclidean space (see, e.g., Pereyra et al. (2016) and references therein). It is also important to extend such studies to the case where the feasible set forms a Riemannian manifold M. Thus, the problem we deal with in this section is of the following form.

Problem 6.1

$$\text{minimize} \quad f(x) := \frac{1}{N} \sum_{i=1}^{N} f_i(x)$$

$$\text{subject to} \quad x \in M. \tag{6.1}$$

6.1.1   Riemannian Stochastic Gradient Descent Method

The *stochastic gradient descent (SGD)* method is one of the simplest stochastic optimization methods. Since the key idea of this algorithm is common to both the Euclidean and Riemannian cases, we introduce the Riemannian version, which includes the Euclidean version as a special case.

We assume that each component f_i, $i = 1, 2, \ldots, N$ is smooth in Problem 6.1. Based on the linearity of the operation grad, we have grad $f(x) = \frac{1}{N} \sum_{i=1}^{N}$ grad $f_i(x)$. Because the steepest descent method requires the computation of grad $f(x_k)$, i.e., all grad $f_i(x_k)$ for $i = 1, 2, \ldots, N$ with N being large, in the kth iteration, it is computationally expensive. The idea of SGD is to randomly choose a part of data, i.e., some components of f, and compute the gradients only for the chosen components.

The stochastic gradient, i.e., grad f_{i_k}, has a property wherein its expectation is equal to the full gradient: $\mathbb{E}\left[\text{grad } f_{i_k}(x)\right] = \frac{1}{N} \sum_{i=1}^{N}$ grad $f_i(x) = $ grad $f(x)$ for any $x \in M$. Each step length t_k, which is also called *learning rate* in the context of machine learning, is usually chosen such that the sequence $\{t_k\}$ satisfies

$$\sum_{k=0}^{\infty} t_k^2 < \infty, \quad \sum_{k=0}^{\infty} t_k = \infty. \tag{6.2}$$

The Euclidean version of the above algorithm can be found in Robbins and Monro (1951). Bonnabel (2013) successfully generalized the Euclidean algorithm to the manifold case, together with convergence analysis, where M is a smooth connected Riemannian manifold.

6.1.2 Riemannian Stochastic Variance Reduced Gradient Method

The steepest descent method is important, even though it is not very fast, since it provides a key idea for the development of other optimization algorithms. Similarly, SGD plays an important role in developing other stochastic optimization algorithms.

Johnson and Zhang (2013) proposed a variance reduction approach, called the *stochastic variance reduced gradient (SVRG) method* to modify the standard Euclidean SGD method. The idea of SVRG is further generalized into the Riemannian case in Zhang et al. (2016) (with the exponential map and parallel translation as a retraction and vector transport, respectively) and Sato et al. (2019) (with a more general retraction and vector transport) together with some convergence analyses. Since SVRG is more complicated than SGD, we first introduce the Euclidean version proposed in Johnson and Zhang (2013).

A key idea of SVRG is, as its name implies, reducing the variance of the stochastic approximation of the gradients by some modification. Consider the Euclidean case, i.e., $M = \mathbb{R}^n$. In the previous subsection, we observed that the expectation of $\nabla f_{i_k}(x)$ is equal to $\nabla f(x)$, where $i_k \in \{1, 2, \dots, N\}$ is randomly chosen in the kth iteration. However, the randomness in choosing f_{i_k} leads to the variance of $\nabla f_{i_k}(x)$, which compels us to choose decaying step lengths as in (6.2). To avoid this situation, the SVRG method uses information from the full gradient $\nabla f(x)$ at a small number of arrived points. Specifically, in SVRG, we regularly keep an obtained point, say $\tilde{x}$, in the iterative process and compute the full gradient $\nabla f(\tilde{x})$. While keeping $\tilde{x}$, we update x_k to x_{k+1} with a randomly chosen index i_k, as in the SGD method, by

$$x_{k+1} = x_k - t_k \big(\nabla f_{i_k}(x_k) - (\nabla f_{i_k}(\tilde{x}) - \nabla f(\tilde{x}))\big). \tag{6.3}$$

The point $\tilde{x}$ is regularly updated in the procedure of SVRG. Compared with $\nabla f_{i_k}(x_k)$ used in the SGD method, the SVRG method uses $\nabla f_{i_k}(x_k) - \big(\nabla f_{i_k}(\tilde{x}) - \nabla f(\tilde{x})\big)$ as a stochastic approximation of $\nabla f(x_k)$. In other words, SVRG modifies $\nabla f_{i_k}(x_k)$ using the additional terms $\nabla f_{i_k}(\tilde{x}) - \nabla f(\tilde{x})$. The expectation of the modified direction is, similar to the case with the stochastic gradient $\nabla f_{i_k}(x_k)$, equal to $\nabla f(x_k)$, but its variance is reduced from that of $\nabla f_{i_k}(x_k)$ because it uses information from the full gradient ∇f. Roughly speaking, this variance reduction technique allows the SVRG method to use a fixed step length and enjoy a faster convergence rate than that of the SGD method. Simultaneously, since we compute the full gradient of f at only some points, the computational cost is lower than that of the steepest descent method.

As we extend this method to a Riemannian version, which we call *Riemannian SVRG (R-SVRG)*, we encounter two familiar issues we discussed in the previous chapters: First, in general, we cannot add x_k and some direction as a tangent vector at x_k on a manifold M. Therefore, we use a retraction R on M for updating $x_k \in M$ once the direction is computed. Second, regarding the direction $\nabla f_{i_k}(x_k) - \big(\nabla f_{i_k}(\tilde{x}) - \nabla f(\tilde{x})\big)$ in (6.3), when we consider the corresponding Riemannian gradients of f, $f_{i_k} : M \to \mathbb{R}$ at $x_k, \tilde{x} \in M$, grad $f_{i_k}(x_k)$ belongs to $T_{x_k}M$, whereas grad $f_{i_k}(\tilde{x})$ and grad $f(\tilde{x})$ belong to $T_{\tilde{x}}M$, and we cannot add grad $f_{i_k}(x_k)$ and $-\big(\text{grad } f_{i_k}(\tilde{x}) - \text{grad } f(\tilde{x})\big)$ together. This situation is similar to the Riemannian conjugate gradient method discussed in Chap. 4. This issue can be resolved by using a vector transport $\mathcal{T}$ on M, i.e., $\mathcal{T}_{\tilde{\eta}_k}\big(\text{grad } f_{i_k}(\tilde{x}) - \text{grad } f(\tilde{x})\big)$ belongs to $T_{x_k}M$, where $\tilde{\eta}_k \in T_{\tilde{x}}M$ satisfies $R_{\tilde{x}}(\tilde{\eta}_k) = x_k$ and R is the retraction associated with $\mathcal{T}$.

As implied by the above discussion, the (R-)SVRG method has a double loop structure, i.e., updating x_k in the inner loop and computing and keeping $\tilde{x}$ in the outer loop. To summarize, the R-SVRG algorithm (Sato et al. 2019) is given in Algorithm 6.1, where s and t denote the counters for the outer and inner loops, respectively, and the step lengths are denoted by α.

Let us take a closer look at Algorithm 6.1. In the sth iteration of the outer loop, we start the iterations of the inner loop with keeping $\tilde{x}^{s-1} \in M$. After m_s iterations of the inner loop, we update $\tilde{x}^{s-1}$ with $\tilde{x}^s \in M$, where $m_s > 0$ is the update frequency of $\tilde{x}$ depending on s. The point $x_{m_s}^s$, which is obtained after m_s iterations of the inner loop, is selected as $\tilde{x}^s$ for the next iteration of the outer loop. Other options for $\tilde{x}^s$ are also proposed in Sato et al. (2019), Zhang et al. (2016). Note that if $R_{\tilde{x}^{s-1}}$ is invertible, then $\tilde{\eta}_{t-1}^s = R_{\tilde{x}^{s-1}}^{-1}(x_{t-1}^s)$ in Algorithm 6.1. However, even if the inverse of $R_{\tilde{x}^{s-1}}$ is not available, it is sufficient to find $\tilde{\eta}_{t-1}^s$ satisfying $R_{\tilde{x}^{s-1}}(\tilde{\eta}_{t-1}^s) = x_{t-1}^s$.

Algorithm 6.1 Riemannian SVRG for Problem 6.1

Input: Riemannian manifold M, retraction R, vector transport $\mathcal{T}$, objective function f on M of the form in (6.1), update frequency $m_s > 0$, and initial point $\tilde{x}^0 \in M$.
Output: Sequence $\{\tilde{x}^s\} \subset M$.
1: **for** $s = 1, 2, \ldots$ **do**
2:    Compute the full Riemannian gradient grad $f(\tilde{x}^{s-1})$.
3:    Store $x_0^s = \tilde{x}^{s-1}$.
4:    **for** $t = 1, 2, \ldots, m_s$ **do**
5:        Choose $i_t^s \in \{1, 2, \ldots, N\}$ uniformly at random.
6:        Compute the tangent vector $\tilde{\eta}_{t-1}^s$ from $\tilde{x}^{s-1}$ to x_{t-1}^s satisfying $R_{\tilde{x}^{s-1}}(\tilde{\eta}_{t-1}^s) = x_{t-1}^s$.
7:        Compute $\xi_t^s = \text{grad } f_{i_t^s}(x_{t-1}^s) - \mathcal{T}_{\tilde{\eta}_{t-1}^s}\big(\text{grad } f_{i_t^s}(\tilde{x}^{s-1}) - \text{grad } f(\tilde{x}^{s-1})\big)$.
8:        Compute a step length α_{t-1}^s.
9:        Update x_t^s from x_{t-1}^s as $x_t^s = R_{x_{t-1}^s}(-\alpha_{t-1}^s \xi_t^s)$.
10:   **end for**
11:   Set $\tilde{x}^s = x_{m_s}^s$.
12: **end for**

Additionally, the R-SVRG method has "a (locally) linear convergence property" with a fixed step length under some conditions, where the concept of "(local) convexity" (see Sect. 6.3.3) of each f_i in (6.1) plays an important role. See Sato et al. (2019), Zhang et al. (2016) for more details.

Remark 6.1 Euclidean stochastic optimization algorithms have been extensively studied, and new algorithms such as the stochastic quasi-Newton (SQN) method (Byrd et al. 2016) and its variant with a variance reduction technique (Moritz et al. 2016), the stochastic recursive gradient algorithm (SARAH) (Nguyen et al. 2017), and the stochastic path integrated differential estimator (SPIDER) (Fang et al. 2018) have been proposed. Accordingly, the Riemannian counterparts of some methods have also been proposed and analyzed, such as R-SQN-VR (Kasai et al. 2018b), R-SRG (Kasai et al. 2018a), and R-SPIDER (Zhou et al. 2021). See also Hosseini and Sra (2020b) for recent studies on stochastic Riemannian optimization.

In machine learning, deep learning has attracted significant attention. A Riemannian optimization library for deep learning, called McTorch, is available (Meghwanshi et al. 2018), which extends PyTorch (Paszke et al. 2017), a Python-based deep learning library. McTorch covers some Riemannian stochastic optimization methods so that one can perform deep learning experiments using PyTorch functions along with Riemannian optimization methods.

6.2 Constrained Optimization on Manifolds

Although this book mainly focuses on unconstrained optimization on manifolds, some studies have also been conducted on constrained optimization on manifolds. In this section, we briefly review some existing results.

A general constrained Euclidean optimization problem is written as follows.

Problem 6.2

$$
\begin{aligned}
\text{minimize} \quad & f(x) \\
\text{subject to} \quad & g_i(x) \le 0, \quad i = 1, 2, \ldots, m, \\
& h_j(x) = 0, \quad j = 1, 2, \ldots, r, \\
& x \in \mathbb{R}^n.
\end{aligned}
$$

Here, $f, g_i, h_j \colon \mathbb{R}^n \to \mathbb{R}$ are real-valued functions.

In the previous chapters, we dealt with special types of equality-constrained optimization problems whose feasible sets form Riemannian manifolds. For example, consider a Euclidean constrained optimization problem to minimize $f(x) := x^T A x$ for given $A \in \mathrm{Sym}(n)$ subject to $h_1(x) := x^T x - 1 = 0$, where $x \in \mathbb{R}^n$. We can find a nice structure of the feasible set $\{x \in \mathbb{R}^n \mid x^T x - 1 = 0\}$ as a Riemannian manifold, i.e., the sphere S^{n-1}, and deal with the problem as an unconstrained Riemannian optimization problem on the sphere S^{n-1} with the formulation of Problem 5.1.

However, in the Riemannian case, when there are some additional equality and/or inequality constraints in addition to the manifold constraint, it may be difficult to handle the entire set of constraints within the setting of unconstrained optimization problems on manifolds. It is then natural to consider a *constrained* optimization problem on a manifold M as follows.

Problem 6.3

$$\begin{aligned}
\text{minimize} \quad & f(x) \\
\text{subject to} \quad & g_i(x) \leq 0, \quad i = 1, 2, \ldots, m, \\
& h_j(x) = 0, \quad j = 1, 2, \ldots, r, \\
& x \in M.
\end{aligned}$$

Here, $f, g_i, h_j \colon M \to \mathbb{R}$ are real-valued functions.

In what follows, we assume that f, g_i, h_j in Problem 6.3 are sufficiently smooth.

For example, we can consider Problem 6.3 with only inequality constraints where $M := S^{n-1}$, $f(x) := x^T A x$ with $A \in \mathrm{Sym}(n)$ given, and $g_i(x) := -x_i$ for $i = 1, 2, \ldots, n$. Another example is the case with $M := \mathrm{St}(p, n)$, $f(X) := \mathrm{tr}(X^T A X)$ with $A \in \mathrm{Sym}(n)$ given, the inequality constraints $X_{kl} \geq 0$ for $k = 1, 2, \ldots, n$ and $l = 1, 2, \ldots, p$ (hence, $m = np$), and the equality constraints $\sum_{k=1}^{n}(XX^T)_{jk} = 1$ for $j = 1, 2, \ldots, n$ (hence, $r = n$). The two examples can be applied to nonnegative principal component analysis and k-means clustering, respectively (Liu and Boumal 2020).

From a theoretical perspective, the *Karush–Kuhn–Tucker (KKT) conditions*, which under some *constraint qualifications (CQs)* are well known as first-order necessary optimality conditions for constrained optimization problems in Euclidean spaces, are generalized to the manifold case (Bergmann and Herzog 2019; Udrişte 1994; Yang et al. 2014). In particular, Bergmann and Herzog (2019) proposed KKT conditions without endowing M with a Riemannian metric, i.e., M in Problem 6.3 is a smooth manifold but is not assumed to be a Riemannian manifold. The *Lagrangian* associated with Problem 6.3 is defined as

$$L(x, \mu, \lambda) := f(x) + \sum_{i=1}^{m} \mu_i g_i(x) + \sum_{j=1}^{r} \lambda_j h_j(x),$$

where $\mu = (\mu_i) \in \mathbb{R}^m$ and $\lambda = (\lambda_j) \in \mathbb{R}^r$ are the Lagrange multipliers associated with the inequality and equality constraints, respectively. The KKT conditions for Problem 6.3 are then written as

$$\mathrm{D}f(x) + \sum_{i=1}^{m} \mu_i \mathrm{D}g_i(x) + \sum_{j=1}^{r} \lambda_j \mathrm{D}h_j(x) = 0, \tag{6.4}$$

$$h_j(x) = 0, \quad j = 1, 2, \ldots, r, \tag{6.5}$$

$$\mu_i \geq 0, \quad g_i(x) \leq 0, \quad \mu_i g_i(x) = 0, \quad i = 1, 2, \ldots, m. \tag{6.6}$$

When M is a Euclidean space, in (6.4), the derivatives are usually replaced with the corresponding Euclidean gradients, e.g., $\mathrm{D}f(x)$ is replaced with $\nabla f(x)$. Similarly, when M is a Riemannian manifold, the derivatives in (6.4) can be replaced with the Riemannian gradients. For the Euclidean case, various types of CQs are known (Bertsekas 2016). In Bergmann and Herzog (2019), the CQs of linear independence (LICQ), Mangasarian–Fromovitz (MFCQ), Abadie (ACQ), and Guignard (GCQ) are generalized to the manifold case, and it is shown that

$$\text{LICQ} \implies \text{MFCQ} \implies \text{ACQ} \implies \text{GCQ} \tag{6.7}$$

holds true; although we omit the detailed description of the CQs here. It is also shown in Bergmann and Herzog (2019) that, if $x_* \in M$ is a local optimal solution to Problem 6.3 and the GCQ, which is the weakest CQ among (6.7), holds at x_*, then there exist $\mu \in \mathbb{R}^m$ and $\lambda \in \mathbb{R}^r$ such that the KKT conditions (6.4)–(6.6) with $x = x_*$ hold.

For an algorithmic aspect, Liu and Boumal (2020) generalized the *augmented Lagrange method (ALM)* for solving Problem 6.2 in Euclidean spaces (Gill and Robinson 2012) to the Riemannian case, which is called the *Riemannian ALM (RALM)*. As usual, we endow M in Problem 6.3 with a Riemannian metric. The augmented Lagrangian for Problem 6.3 is defined in the same manner as in the Euclidean case by

$$L_\rho(x, \mu, \lambda)$$
$$:= f(x) + \frac{\rho}{2} \left(\sum_{j=1}^{r} \left(h_j(x) + \frac{\lambda_j}{\rho} \right)^2 + \sum_{i=1}^{m} \left(\max \left\{ 0, \frac{\mu_i}{\rho} + g_i(x) \right\} \right)^2 \right),$$

where $\mu = (\mu_i) \in \mathbb{R}^m$ with $\mu_i \geq 0$ for $i = 1, 2, \ldots, m$, $\lambda = (\lambda_j) \in \mathbb{R}^r$, and $\rho > 0$ is a penalty parameter. The RALM alternates between minimizing L for fixed (λ, μ) to update x and updating (λ, μ). The exact penalty method for Problem 6.3 is also discussed in Liu and Boumal (2020).

6.3 Other Emerging Methods and Related Topics

This book mainly focuses on smooth unconstrained optimization problems on Riemannian manifolds and first-order optimization algorithms for solving them. In this section, we briefly review some topics, including the emerging methods that have not been examined in this book, with references for interested readers.

6.3.1 Second-Order Methods

The steepest descent and conjugate gradient methods use the gradient of the objective function f and are called first-order methods. In contrast, Newton's method uses the information of the Hessian, i.e., second-order information of f.

On a Riemannian manifold M, the Hessian of the objective function f at $x_k \in M$, denoted by Hess $f(x_k)$, is a generalization of the usual Hessian matrix in $\mathbb{R}^n$; however, we do not formally define Hess f in this book. Here, we simply emphasize that Hess $f(x_k)$ is a linear operator from $T_{x_k}M$ to $T_{x_k}M$.

Although we do not deal with it extensively, the algorithm for *Riemannian Newton's method* is similar to that of the Euclidean version and is expressed in the framework of Algorithm 3.2 (see, e.g., Absil et al. (2008)). The search direction $\eta_k \in T_{x_k}M$ is computed as a solution to Newton's equation

$$\text{Hess } f(x_k)[\eta_k] = -\text{grad } f(x_k), \tag{6.8}$$

which is a generalization of the Euclidean version (2.28) and is a linear system with respect to $\eta_k \in T_{x_k}M$.

We recall Krylov subspace methods, such as the linear conjugate gradient method discussed in Sect. 4.1, to solve the linear system (6.8). Krylov subspace methods are not limited to the case of solving $Ax = b$ with a matrix A and column vector b; they can also be used for $\mathscr{A}(x) = b$, where $\mathscr{A} : V \to V$ is a linear operator in a vector space V and $x, b \in V$. Therefore, for given x_k, we can apply Krylov subspace methods where $\mathscr{A} = \text{Hess } f(x_k)$ and $b = -\text{grad } f(x_k)$ to (6.8). For more detail, see Aihara and Sato (2017).

While Riemannian Newton's method is still being developed (Fernandes et al. 2017; Hu et al. 2018), the quasi-Newton method including BFGS-type one, where we approximate (the inverse of) the Hessian of f, is also generalized to the Riemannian case (Huang et al. 2018a, 2015; Ring and Wirth 2012). The trust-region method in Euclidean spaces (Nocedal and Wright 2006) also uses a second-order approximation of the objective function and is generalized to the Riemannian case (Absil et al. 2007, 2008). Further, adaptive regularization with cubics is generalized to the Riemannian case and analyzed (Agarwal et al. 2020; Qi 2011; Zhang and Zhang 2018).

6.3.2 Nonsmooth Riemannian Optimization

Although this book has focused on smooth objective functions thus far, Riemannian optimization problems whose objective functions are nonsmooth are also important. Such nonsmooth Riemannian optimization problems include sparse principal component analysis (Genicot et al. 2015), range-based independent component analysis (Selvan et al. 2013), robust low-rank matrix completion (Cambier et al. 2016), and restoration of manifold-valued images (Bacák et al. 2016; Grohs and Sprecher 2016). These applications, among others, are reviewed in Absil and Hosseini (2019).

An important concept for nonsmooth optimization is the *subgradient* of the objective function, which is a generalization of the gradient and can be defined for nonsmooth functions (Ferreira 2006; Hosseini and Pouryayevali 2011; Hosseini and Uschmajew 2017). The Riemannian versions of the subgradient method have been examined in Borckmans et al. (2014), Dirr et al. (2007). In Grohs and Hosseini (2016), a nonsmooth Riemannian trust-region method is proposed. The work performed in Kovnatsky et al. (2016) focuses on the manifold alternating directions method of multipliers (MADMM), which can be viewed as an extension of classical ADMM (Boyd et al. 2011) to nonsmooth optimization problems on manifolds. The work in Zhang et al. (2020) is also an ADMM-like approach in Riemannian optimization. Proximal algorithms, of which the Euclidean version are reviewed in Parikh and Boyd (2014), are generalized to the Riemannian case, e.g., proximal point method on general Riemannian manifolds (de Carvalho Bento et al. 2016) and proximal gradient method on the Stiefel manifold (Chen et al. 2020).

6.3.3 Geodesic and Retraction Convexity

The notion of *geodesic convexity* of functions on Riemannian manifolds is a generalization of convexity in Euclidean spaces (Rapcsák 1991). More generally, *retraction convexity* has also been introduced in Huang (2013), Huang et al. (2015); a function $f : M \to \mathbb{R}$ on a Riemannian manifold M is said to be retraction convex with respect to a retraction R in a set $S \subset M$ if, for all $x \in M$ and $\eta \in T_x M$ with $\|\eta\|_x = 1$, the function $t \mapsto f(R_x(t\eta))$ is convex (with respect to Definition 2.19) for all t satisfying $R_x(\tau\eta) \in S$ for all $\tau \in [0, t]$.

For geodesically convex optimization, complexity analyses for some algorithms are performed in Zhang and Sra (2016) and accelerated first-order methods are proposed in Liu et al. (2017). There are further studies ongoing and more development is expected.

6.3.4 Multi-objective Optimization on Riemannian Manifolds

In some applications, one may wish to minimize or maximize multiple objective functions. Such problems are called multi-objective optimization problems, which have also been studied extensively (Eichfelder 2009; Evans 1984; Fliege and Svaiter 2000; Fukuda and Drummond 2014). However, many basic concepts of multi-objective optimization originate from the studies of single-objective optimization. Although this book only dealt with single-objective optimization, our discussion can be extended to multi-objective optimization on Riemannian manifolds with appropriate generalization. See, e.g., Bento et al. (2012, 2018), Ferreira et al. (2020).

References

Absil PA, Baker CG, Gallivan KA (2007) Trust-region methods on Riemannian manifolds. Found Comput Math 7(3):303–330

Absil PA, Gallivan KA (2006) Joint diagonalization on the oblique manifold for independent component analysis. In: Proceedings of the 2006 IEEE International Conference on Acoustics, Speech and Signal Processing, pp V-945–V-948

Absil PA, Hosseini S (2019) A collection of nonsmooth Riemannian optimization problems. In: Nonsmooth Optimization and Its Applications. Birkhäuser, Cham, pp 1–15

Absil PA, Mahony R, Sepulchre R (2008) Optimization Algorithms on Matrix Manifolds. Princeton University Press, Princeton

Absil PA, Malick J (2012) Projection-like retractions on matrix manifolds. SIAM J Optim 22(1):135–158

Adler RL, Dedieu JP, Margulies JY, Martens M, Shub M (2002) Newton's method on Riemannian manifolds and a geometric model for the human spine. IMA J Numer Anal 22(3):359–390

Afsari B, Krishnaprasad PS (2004) Some gradient based joint diagonalization methods for ICA. In: Independent Component Analysis and Blind Signal Separation. Springer, Heidelberg, pp 437–444

Agarwal N, Boumal N, Bullins B, Cartis C (2020) Adaptive regularization with cubics on manifolds. Math Program. https://doi.org/10.1007/s10107-020-01505-1

Aihara K, Sato H (2017) A matrix-free implementation of Riemannian Newton's method on the Stiefel manifold. Optim Lett 11(8):1729–1741

Al-Baali M (1985) Descent property and global convergence of the Fletcher-Reeves method with inexact line search. IMA J Numer Anal 5(1):121–124

Axler S (2015) Linear Algebra Done Right, 3rd edn. Springer, Cham

Bačák M, Bergmann R, Steidl G, Weinmann A (2016) A second order nonsmooth variational model for restoring manifold-valued images. SIAM J Sci Comput 38(1):A567–A597

Bento GC, Ferreira OP, Oliveira PR (2012) Unconstrained steepest descent method for multicriteria optimization on Riemannian manifolds. J Optim Theory Appl 154(1):88–107

Bento GdeC, da Cruz Neto JX, de Meireles LV (2018) Proximal point method for locally Lipschitz functions in multiobjective optimization of Hadamard manifolds. J Optim Theory Appl 179(1):37–52

Benzi M (2002) Preconditioning techniques for large linear systems: a survey. J Comput Phys 182(2):418–477

© Springer Nature Switzerland AG 2021
H. Sato, *Riemannian Optimization and Its Applications*,
SpringerBriefs in Control, Automation and Robotics,
https://doi.org/10.1007/978-3-030-62391-3

Bergmann R (2019) manopt.jl — Optimization on Manifolds in Julia. http://www.manoptjl.org

Bergmann R, Herzog R (2019) Intrinsic formulation of KKT conditions and constraint qualifications on smooth manifolds. SIAM J Optim 29(4):2423–2444

Bertsekas DP (2016) Nonlinear Programming, 3rd edn. Athena Scientific, Belmont

Bhatia R (2007) Positive Definite Matrices. Princeton University Press, Princeton

Bodmann BG, Haas J (2015) Frame potentials and the geometry of frames. J Fourier Anal Appl 21(6):1344–1383

Bonnabel S (2013) Stochastic gradient descent on Riemannian manifolds. IEEE Trans Autom Control 58(9):2217–2229

Borckmans PB, Selvan SE, Boumal N, Absil PA (2014) A Riemannian subgradient algorithm for economic dispatch with valve-point effect. J Comput Appl Math 255:848–866

Boumal N, Mishra B, Absil PA, Sepulchre R (2014) Manopt, a Matlab toolbox for optimization on manifolds. J Mach Learn Res 15(1):1455–1459

Boyd S, Parikh N, Chu E, Peleato B, Eckstein J (2011) Distributed optimization and statistical learning via the alternating direction method of multipliers. Found Trends Mach Learn 3(1):1–122

Boyd S, Vandenberghe L (2004) Convex Optimization. Cambridge University Press, Cambridge

Brockett RW (1993) Differential geometry and the design of gradient algorithms. Proc Symp Pure Math 54:69–92

Byrd RH, Hansen SL, Nocedal J, Singer Y (2016) A stochastic quasi-Newton method for large-scale optimization. SIAM J Optim 26(2):1008–1031

Cambier L, Absil PA (2016) Robust low-rank matrix completion by Riemannian optimization. SIAM J Sci Comput 38(5):S440–S460

Carlson D, Hsieh YP, Collins E, Carin L, Cevher V (2015) Stochastic spectral descent for discrete graphical models. IEEE J Sel Top Signal Process 10(2):296–311

de Carvalho Bento G, da Cruz Neto JX, Oliveira PR (2016) A new approach to the proximal point method: convergence on general Riemannian manifolds. J Optim Theory Appl 168(3):743–755

Chen S, Ma S, So MCA, Zhang T (2020) Proximal gradient method for nonsmooth optimization over the Stiefel manifold. SIAM J Optim 30(1):210–239

Congedo M, Phlypo R, Pham DT (2011) Approximate joint singular value decomposition of an asymmetric rectangular matrix set. IEEE Trans Signal Process 59(1):415–424

Dai YH, Yuan Y (1999) A nonlinear conjugate gradient method with a strong global convergence property. SIAM J Optim 10(1):177–182

Dai YH, Yuan Y (2001) A three-parameter family of nonlinear conjugate gradient methods. Math Comput 70(235):1155–1167

Dirr G, Helmke U, Lageman C (2007) Nonsmooth Riemannian optimization with applications to sphere packing and grasping. In: Lagrangian and Hamiltonian Methods for Nonlinear Control 2006. Springer, Heidelberg, pp 29–45

Dullerud GE, Paganini F (2000) A Course in Robust Control Theory: A Convex Approach. Springer, New York

Dummit DS, Foote RM (2003) Abstract Algebra, 3rd edn. Wiley, Hoboken

Edelman A, Arias TA, Smith ST (1998) The geometry of algorithms with orthogonality constraints. SIAM J Matrix Anal Appl 20(2):303–353

Edelman A, Smith ST (1996) On conjugate gradient-like methods for eigen-like problems. BIT Numer Math 36(3):494–508

Eichfelder G (2009) An adaptive scalarization method in multiobjective optimization. SIAM J Optim 19(4):1694–1718

Eisenstat SC, Elman HC, Schultz MH (1983) Variational iterative methods for nonsymmetric systems of linear equations. SIAM J Numer Anal 20(2):345–357

Evans GW (1984) An overview of techniques for solving multiobjective mathematical programs. Manage Sci 30(11):1268–1282

Fang C, Li CJ, Lin Z, Zhang T (2018) SPIDER: Near-optimal non-convex optimization via stochastic path-integrated differential estimator. Adv Neural Inf Process Syst 31:689–699

Fernandes TA, Ferreira OP, Yuan J (2017) On the superlinear convergence of Newton's method on Riemannian manifolds. J Optim Theory Appl 173(3):828–843

Ferreira OP (2006) Proximal subgradient and a characterization of Lipschitz function on Riemannian manifolds. J Math Anal Appl 313(2):587–597

Ferreira OP, Louzeiro MS, Prudente LF (2020) Iteration-complexity and asymptotic analysis of steepest descent method for multiobjective optimization on Riemannian manifolds. J Optim Theory Appl 184(2):507–533

Fletcher R (1976) Conjugate gradient methods for indefinite systems. In: Numerical Analysis. Springer, Heidelberg, pp 73–89

Fletcher R (2000) Practical Methods of Optimization, 2nd edn. Wiley, Hoboken

Fletcher R, Reeves CM (1964) Function minimization by conjugate gradients. Comput J 7(2):149–154

Fliege J, Svaiter BF (2000) Steepest descent methods for multicriteria optimization. Math Methods Oper Res 51(3):479–494

Forst W, Hoffmann D (2010) Optimization: Theory and Practice. Springer, New York

Fukuda EH, Drummond LMG (2014) A survey on multiobjective descent methods. Pesquisa Operacional 34(3):585–620

Gabay D (1982) Minimizing a differentiable function over a differential manifold. J Optim Theory Appl 37(2):177–219

Genicot M, Huang W, Trendafilov NT (2015) Weakly correlated sparse components with nearly orthonormal loadings. In: International Conference on Geometric Science of Information. Springer, Cham, pp 484–490

Gill PE, Robinson DP (2012) A primal-dual augmented Lagrangian. Comput Optim Appl 51(1):1–25

Golub GH, Van Loan CF (2013) Matrix Computations, 4th edn. Johns Hopkins University Press, Baltimore

Grasedyck L, Kressner D, Tobler C (2013) A literature survey of low-rank tensor approximation techniques. GAMM-Mitteilungen 36(1):53–78

Grohs P, Hosseini S (2016) Nonsmooth trust region algorithms for locally Lipschitz functions on Riemannian manifolds. IMA J Numer Anal 36(3):1167–1192

Grohs P, Sprecher M (2016) Total variation regularization on Riemannian manifolds by iteratively reweighted minimization. Inf Infer J IMA 5(4):353–378

Gugercin S, Antoulas AC, Beattie C (2008) $\mathcal{H}_2$ model reduction for large-scale linear dynamical systems. SIAM J Matrix Anal Appl 30(2):609–638

Hager WW, Zhang H (2005) A new conjugate gradient method with guaranteed descent and an efficient line search. SIAM J Optim 16(1):170–192

Hager WW, Zhang H (2006) A survey of nonlinear conjugate gradient methods. Pac J Optim 2(1):35–58

Hastie T, Tibshirani R, Wainwright M (2015) Statistical Learning with Sparsity: The Lasso and Generalizations. CRC Press, Boca Raton

He P, Xu X, Ding J, Fan B (2020) Low-rank nonnegative matrix factorization on Stiefel manifold. Inf Sci 514:131–148

Helmke U, Moore JB (1994) Optimization and Dynamical Systems. Communications and Control Engineering Series, Springer-Verlag, (London and New York)

Hestenes MR, Stiefel E (1952) Methods of conjugate gradients for solving linear systems. J Res Natl Bur Stand 49(6):409–436

Hori G (2010) Joint SVD and its application to factorization method. In: International Conference on Latent Variable Analysis and Signal Separation. Springer, Heidelberg, pp 563–570

Hosseini R, Sra S (2015) Matrix manifold optimization for Gaussian mixtures. Adv Neural Inf Process Syst 28:910–918

Hosseini R, Sra S (2020) An alternative to EM for Gaussian mixture models: batch and stochastic Riemannian optimization. Math Program 181(1):187–223

Hosseini R, Sra S (2020) Recent advances in stochastic Riemannian optimization. In: Handbook of Variational Methods for Nonlinear Geometric Data. Springer, Cham, pp 527–554

Hosseini S, Pouryayevali MR (2011) Generalized gradients and characterization of epi-Lipschitz sets in Riemannian manifolds. Nonlinear Anal Theory Methods Appl 74(12):3884–3895

Hosseini S, Uschmajew A (2017) A Riemannian gradient sampling algorithm for nonsmooth optimization on manifolds. SIAM J Optim 27(1):173–189

Hu J, Milzarek A, Wen Z, Yuan Y (2018) Adaptive quadratically regularized Newton method for Riemannian optimization. SIAM J Matrix Anal Appl 39(3):1181–1207

Huang W (2013) Optimization Algorithms on Riemannian Manifolds with Applications. Ph. D. Thesis, Florida State University

Huang W, Absil PA, Gallivan KA (2017) Intrinsic representation of tangent vectors and vector transports on matrix manifolds. Numerische Mathematik 136(2):523–543

Huang W, Absil PA, Gallivan KA (2018) A Riemannian BFGS method without differentiated retraction for nonconvex optimization problems. SIAM J Optim 28(1):470–495

Huang W, Absil PA, Gallivan KA, Hand P (2018) ROPTLIB: an object-oriented C++ library for optimization on Riemannian manifolds. ACM Trans Math Softw 44(4):43:1–43:21

Huang W, Gallivan KA, Absil PA (2015) A Broyden class of quasi-Newton methods for Riemannian optimization. SIAM J Optim 25(3):1660–1685

Ishizaki T, Sandberg H, Kashima K, Imura J, Aihara K (2015) Dissipativity-preserving model reduction for large-scale distributed control systems. IEEE Trans Autom Control 60(4):1023–1037

Johnson R, Zhang T (2013) Accelerating stochastic gradient descent using predictive variance reduction. Adv Neural Inf Process Syst 26:315–323

Journée M, Bach F, Absil PA, Sepulchre R (2010) Low-rank optimization on the cone of positive semidefinite matrices. SIAM J Optim 20(5):2327–2351

Kasai H, Mishra B (2016) Low-rank tensor completion: a Riemannian manifold preconditioning approach. In: Proceedings of the 33rd International Conference on Machine Learning, pp 1012–1021

Kasai H, Sato H, Mishra B (2018) Riemannian stochastic quasi-Newton algorithm with variance reduction and its convergence analysis. In: Proceedings of the 21st International Conference on Artificial Intelligence and Statistics, pp 269–278

Kasai H, Sato H, Mishra B (2018) Riemannian stochastic recursive gradient algorithm. In: Proceedings of the 35th International Conference on Machine Learning, pp 2521–2529

Kovnatsky A, Glashoff K, Bronstein MM (2016) MADMM: a generic algorithm for non-smooth optimization on manifolds. In: European Conference on Computer Vision. Springer, Cham, pp 680–696

Kressner D, Steinlechner M, Vandereycken B (2014) Low-rank tensor completion by Riemannian optimization. BIT Numer Math 54(2):447–468

Lee DD, Seung HS (1999) Learning the parts of objects by non-negative matrix factorization. Nature 401(6755):788–791

Lee JM (2012) Introduction to Smooth Manifolds, 2nd edn. Springer, New York

Lee JM (1997) Riemannian Manifolds: An Introduction to Curvature. Springer, New York

Lee T, Park FC (2018) A geometric algorithm for robust multibody inertial parameter identification. IEEE Robot Autom Lett 3(3):2455–2462

Lemaréchal C (1981) A view of line-searches. In: Optimization and Optimal Control. Springer, Heidelberg, pp 59–78

Liu C, Boumal N (2020) Simple algorithms for optimization on Riemannian manifolds with constraints. Appl Math Optim 82(3):949–981

Liu Y, Shang F, Cheng J, Cheng H, Jiao L (2017) Accelerated first-order methods for geodesically convex optimization on Riemannian manifolds. Adv Neural Inf Process Syst 30:4868–4877

Liu Y, Storey C (1991) Efficient generalized conjugate gradient algorithms, part 1: Theory. J Optim Theory Appl 69(1):129–137

Luenberger DG, Ye Y (2008) Linear and Nonlinear Programming, 3rd edn. Springer, Heidelberg

Martin S, Raim AM, Huang W, Adragni KP (2016) ManifoldOptim: An R interface to the ROPTLIB library for Riemannian manifold optimization. arXiv preprint arXiv:1612.03930

Massart E, Absil PA (2020) Quotient geometry with simple geodesics for the manifold of fixed-rank positive-semidefinite matrices. SIAM J Matrix Anal Appl 41(1):171–198

Meghwanshi M, Jawanpuria P, Kunchukuttan A, Kasai H, Mishra B (2018) McTorch, a manifold optimization library for deep learning. arXiv preprint arXiv:1810.01811

Mendelson B (1990) Introduction to Topology: Third Edition, Dover, New York

Meyer CD (2000) Matrix Analysis and Applied Linear Algebra. SIAM, Philadelphia

Mishra B, Meyer G, Bonnabel S, Sepulchre R (2014) Fixed-rank matrix factorizations and Riemannian low-rank optimization. Comput Stat 29(3–4):591–621

Mishra B, Sepulchre R (2016) Riemannian preconditioning. SIAM J Optim 26(1):635–660

Moakher M (2005) A differential geometric approach to the geometric mean of symmetric positive-definite matrices. SIAM J Matrix Anal Appl 26(3):735–747

Moritz P, Nishihara R, Jordan M (2016) A linearly-convergent stochastic L-BFGS algorithm. In: Proceedings of the 19th International Conference on Artificial Intelligence and Statistics, pp 249–258

Munkres J (2000) Topology, 2nd Edition. Pearson, Upper Saddle River

Wolsey LA, Nemhauser GL (1999) Integer and Combinatorial Optimization. Wiley, New York

Nesterov Y (2012) Efficiency of coordinate descent methods on huge-scale optimization problems. SIAM J Optim 22(2):341–362

Nguyen LM, Liu J, Scheinberg K, Takáč M (2017) SARAH: A novel method for machine learning problems using stochastic recursive gradient. In: Proceedings of the 34th International Conference on Machine Learning, pp 2613–2621

Nocedal J, Wright SJ (2006) Numerical Optimization, 2nd edn. Springer, New York

Olshanskii MA, Tyrtyshnikov EE (2014) Iterative Methods for Linear Systems: Theory and Applications. SIAM, Philadelphia

Paige CC, Saunders MA (1975) Solution of sparse indefinite systems of linear equations. SIAM J Numer Anal 12(4):617–629

Parikh N, Boyd S (2014) Proximal algorithms. Found Trends Optim 1(3):127–239

Paszke A, Gross S, Chintala S, Chanan G, Yang E, DeVito Z, Lin Z, Desmaison A, Antiga L, Lerer A (2017) Automatic differentiation in PyTorch. In: NIPS 2017 Workshop on Automatic Differentiation

Pereyra M, Schniter P, Chouzenoux E, Pesquet JC, Tourneret JY, Hero AO, McLaughlin S (2016) A survey of stochastic simulation and optimization methods in signal processing. IEEE J Sel Top Signal Process 10(2):224–241

Pfeffer M, Uschmajew A, Amaro A, Pfeffer U (2019) Data fusion techniques for the integration of multi-domain genomic data from uveal melanoma. Cancers 11(10):1434

Polak E, Ribiére G (1969) Note sur la convergence de méthodes de directions conjuguées. Rev Francaise Informat Recherche Operationelle 3(R1):35–43

Polyak BT (1969) The conjugate gradient method in extremal problems. USSR Comput Math Math Phys 9(4):94–112

Pompili F, Gillis N, Absil PA, Glineur F (2014) Two algorithms for orthogonal nonnegative matrix factorization with application to clustering. Neurocomputing 141:15–25

Qi C (2011) Numerical Optimization Methods on Riemannian Manifolds. Ph. D. Thesis, Florida State University

Rapcsák T (1991) Geodesic convexity in nonlinear optimization. J Optim Theory Appl 69(1):169–183

Rapcsák T (1997) Smooth Nonlinear Optimization in R^n. Springer Science & Business Media, Dordrecht

Ring W, Wirth B (2012) Optimization methods on Riemannian manifolds and their application to shape space. SIAM J Optim 22(2):596–627

Robbins H, Monro S (1951) A stochastic approximation method. Ann Math Stat 22(3):400–407

Ruszczyński A (2006) Nonlinear Optimization. Princeton University Press, Princeton

Saad Y (2003) Iterative Methods for Sparse Linear Systems. SIAM, Philadelphia

Saad Y, Schultz MH (1986) GMRES: A generalized minimal residual algorithm for solving non-symmetric linear systems. SIAM J Sci Stat Comput 7(3):856–869

Sakai T (1996) Riemannian Geometry. American Mathematical Society, Providence

Sakai H, Iiduka H (2020) Hybrid Riemannian conjugate gradient methods with global convergence properties. Comput Optim Appl 77(3):811–830

Sato H (2014) Riemannian conjugate gradient method for complex singular value decomposition problem. In: Proceedings of the 53rd IEEE Conference on Decision and Control. IEEE, pp 5849–5854

Sato H (2015) Joint singular value decomposition algorithm based on the Riemannian trust-region method. JSIAM Letters 7:13–16

Sato H (2016) A Dai–Yuan-type Riemannian conjugate gradient method with the weak Wolfe conditions. Comput Optim Appl 64(1):101–118

Sato H (2017) Riemannian Newton-type methods for joint diagonalization on the Stiefel manifold with application to independent component analysis. Optimization 66(12):2211–2231

Sato H, Aihara K (2019) Cholesky QR-based retraction on the generalized Stiefel manifold. Comput Optim Appl 72(2):293–308

Sato H, Iwai T (2013) A complex singular value decomposition algorithm based on the Riemannian Newton method. In: Proceedings of the 52nd IEEE Conference on Decision and Control. IEEE, pp 2972–2978

Sato H, Iwai T (2013) A Riemannian optimization approach to the matrix singular value decomposition. SIAM J Optim 23(1):188–212

Sato H, Iwai T (2014) Optimization algorithms on the Grassmann manifold with application to matrix eigenvalue problems. Jpn J Ind Appl Math 31(2):355–400

Sato H, Iwai T (2015) A new, globally convergent Riemannian conjugate gradient method. Optimization 64(4):1011–1031

Sato H, Kasai H, Mishra B (2019) Riemannian stochastic variance reduced gradient algorithm with retraction and vector transport. SIAM J Optim 29(2):1444–1472

Sato H, Sato K (2015) Riemannian trust-region methods for H^2 optimal model reduction. In: Proceedings of the 54th IEEE Conference on Decision and Control. IEEE, pp 4648–4655

Sato H, Sato K (2016) A new H^2 optimal model reduction method based on Riemannian conjugate gradient method. In: Proceedings of the 55th IEEE Conference on Decision and Control. IEEE, pp 5762–5768

Sato H, Sato K (2017) Riemannian optimal system identification algorithm for linear MIMO systems. IEEE Control Syst Lett 1(2):376–381

Sato K (2017) Riemannian optimal control and model matching of linear port-Hamiltonian systems. IEEE Trans Autom Control 62(12):6575–6581

Sato K (2019) Riemannian optimal model reduction of stable linear systems. IEEE Access 7:14689–14698

Sato K, Sato H (2018) Structure-preserving H^2 optimal model reduction based on the Riemannian trust-region method. IEEE Trans Autom Control 63(2):505–512

Sato K, Sato H, Damm T (2020) Riemannian optimal identification method for linear systems with symmetric positive-definite matrix. IEEE Trans Autom Control 65(11):4493–4508

Schrijver A (2003) Combinatorial Optimization: Polyhedra and Efficiency. Springer, Berlin

Selvan SE, Borckmans PB, Chattopadhyay A, Absil PA (2013) Spherical mesh adaptive direct search for separating quasi-uncorrelated sources by range-based independent component analysis. Neural Comput 25(9):2486–2522

Shub M (1986) Some remarks on dynamical systems and numerical analysis. In: Dynamical Systems and Partial Differential Equations: Proceedings of VII ELAM, pp 69–92

Smith ST (1994) Optimization techniques on Riemannian manifolds. In: Hamiltonian and Gradient Flows, Algorithms and Control. American Mathematical Soc, pp 113–135

Snyman JA (2005) Practical Mathematical Optimization: An Introduction to Basic Optimization Theory and Classical and New Gradient-Based Algorithms. Springer, Heidelberg

Steinlechner M (2016) Riemannian optimization for high-dimensional tensor completion. SIAM J Sci Comput 38(5):S461–S484

Stiefel E (1955) Relaxationsmethoden bester strategie zur lösung linearer gleichungssysteme. Commentarii Mathematici Helvetici 29(1):157–179

Theis FJ, Cason TP, Absil PA (2009) Soft dimension reduction for ICA by joint diagonalization on the Stiefel manifold. In: Proceedings of the 8th International Conference on Independent Component Analysis and Signal Separation. Springer-Verlag, Berlin, Heidelberg, pp 354–361

Townsend J, Koep N, Weichwald S (2016) Pymanopt: A Python toolbox for optimization on manifolds using automatic differentiation. J Mach Learn Res 17(137):1–5

Traversaro S, Brossette S, Escande A, Nori F (2016) Identification of fully physical consistent inertial parameters using optimization on manifolds. In: Proceedings of the 2016 IEEE/RSJ International Conference on Intelligent Robots and Systems. IEEE, pp 5446–5451

Tu LW (2011) An Introduction to Manifolds, 2nd edn. Springer, New York

Udrişte C (1994) Convex Functions and Optimization Methods on Riemannian Manifolds. Springer Science & Business Media, The Netherlands

Usevich K, Markovsky I (2014) Optimization on a Grassmann manifold with application to system identification. Automatica 50(6):1656–1662

Vandereycken B (2013) Low-rank matrix completion by Riemannian optimization. SIAM J Optim 23(2):1214–1236

Yamada I, Ezaki T (2003) An orthogonal matrix optimization by dual Cayley parametrization technique. In: Proceedings of the 4th International Symposium on Independent Component Analysis and Blind Signal Separation, pp 35–40

Yan WY, Lam J (1999) An approximate approach to H^2 optimal model reduction. IEEE Trans Autom Control 44(7):1341–1358

Yang WH, Zhang LH, Song R (2014) Optimality conditions for the nonlinear programming problems on Riemannian manifolds. Pac J Optim 10(2):415–434

Yao TT, Bai ZJ, Jin XQ, Zhao Z (2020) A geometric Gauss-Newton method for least squares inverse eigenvalue problems. BIT Numer Math 60(3):825–852

Yao TT, Bai ZJ, Zhao Z, Ching WK (2016) A Riemannian Fletcher–Reeves conjugate gradient method for doubly stochastic inverse eigenvalue problems. SIAM J Matrix Anal Appl 37(1):215–234

Yger F, Berar M, Gasso G, Rakotomamonjy A (2012) Adaptive canonical correlation analysis based on matrix manifolds. In: Proceedings of the 29th International Conference on Machine Learning, pp 1071–1078

Zhang H, Reddi SJ, Sra S (2016) Riemannian SVRG: Fast stochastic optimization on Riemannian manifolds. Adv Neural Inf Process Syst 29:4592–4600

Zhang H, Sra S (2016) First-order methods for geodesically convex optimization. In: Proceedings of the 29th Annual Conference on Learning Theory, pp 1617–1638

Zhang J, Ma S, Zhang S (2020) Primal-dual optimization algorithms over Riemannian manifolds: an iteration complexity analysis. Math Program 184(1–2):445–490

Zhang J, Zhang S (2018) A cubic regularized Newton's method over Riemannian manifolds. arXiv preprint arXiv:1805.05565

Zhao Z, Bai ZJ, Jin XQ (2015) A Riemannian Newton algorithm for nonlinear eigenvalue problems. SIAM J Matrix Anal Appl 36(2):752–774

Zhou P, Yuan X, Yan S, Feng J (2021) Faster first-order methods for stochastic non-convex optimization on Riemannian manifolds. IEEE Trans Pattern Anal Mach Intell 43(2):459–472

Zhu X (2017) A Riemannian conjugate gradient method for optimization on the Stiefel manifold. Comput Optim Appl 67(1):73–110

Zhu X, Sato H (2020) Riemannian conjugate gradient methods with inverse retraction. Comput Optim Appl 77(3):779–810

Series Editors' Biographies

Tamer Başar is with the University of Illinois at Urbana-Champaign, where he holds the academic positions of Swanlund Endowed Chair, Center for Advanced Study (CAS) Professor of Electrical and Computer Engineering, Professor at the Coordinated Science Laboratory, Professor at the Information Trust Institute, and Affiliate Professor of Mechanical Science and Engineering. He is also the Director of the Center for Advanced Study—a position he has been holding since 2014. At Illinois, he has also served as Interim Dean of Engineering (2018) and Interim Director of the Beckman Institute for Advanced Science and Technology (2008–2010). He received the B.S.E.E. degree from Robert College, Istanbul, and the M.S., M.Phil., and Ph.D. degrees from Yale University. He has published extensively in systems, control, communications, networks, optimization, learning, and dynamic games, including books on non-cooperative dynamic game theory, robust control, network security, wireless and communication networks, and stochastic networks, and has current research interests that address fundamental issues in these areas along with applications in multi-agent systems, energy systems, social networks, cyber-physical systems, and pricing in networks.

In addition to his editorial involvement with these Briefs, Başar is also the Editor of two Birkhäuser series on *Systems & Control: Foundations & Applications and Static & Dynamic Game Theory: Foundations & Applications*, the Managing Editor of the *Annals of the International Society of Dynamic Games* (ISDG), and member of editorial and advisory boards of several international journals in control, wireless networks, and applied mathematics. Notably, he was also the Editor-in-Chief of *Automatica* between 2004 and 2014. He has received several awards and recognitions over the years, among which are the Medal of Science of Turkey (1993); Bode Lecture Prize (2004) of IEEE CSS; Quazza Medal (2005) of IFAC; Bellman Control Heritage Award (2006) of AACC; Isaacs Award (2010) of ISDG; Control Systems Technical Field Award of IEEE (2014); and a number of international honorary doctorates and professorships. He is a member of the US National Academy of Engineering, a Life Fellow of IEEE, Fellow of IFAC, and Fellow of SIAM. He has served as an IFAC Advisor (2017–), a Council Member of IFAC (2011–2014), president of

© Springer Nature Switzerland AG 2021
H. Sato, *Riemannian Optimization and Its Applications*,
SpringerBriefs in Control, Automation and Robotics,
https://doi.org/10.1007/978-3-030-62391-3

AACC (2010–2011), president of CSS (2000), and founding president of ISDG (1990–1994).

Miroslav Krstic is Distinguished Professor of Mechanical and Aerospace Engineering, holds the Alspach endowed chair, and is the founding director of the Cymer Center for Control Systems and Dynamics at UC San Diego. He also serves as Senior Associate Vice Chancellor for Research at UCSD. As a graduate student, Krstic won the UC Santa Barbara best dissertation award and student best paper awards at CDC and ACC. Krstic has been elected Fellow of IEEE, IFAC, ASME, SIAM, AAAS, IET (UK), AIAA (Assoc. Fellow), and as a foreign member of the Serbian Academy of Sciences and Arts and of the Academy of Engineering of Serbia. He has received the SIAM Reid Prize, ASME Oldenburger Medal, Nyquist Lecture Prize, Paynter Outstanding Investigator Award, Ragazzini Education Award, IFAC Nonlinear Control Systems Award, Chestnut textbook prize, Control Systems Society Distinguished Member Award, the PECASE, NSF Career, and ONR Young Investigator awards, the Schuck ('96 and '19) and Axelby paper prizes, and the first UCSD Research Award given to an engineer. Krstic has also been awarded the Springer Visiting Professorship at UC Berkeley, the Distinguished Visiting Fellowship of the Royal Academy of Engineering, and the Invitation Fellowship of the Japan Society for the Promotion of Science. He serves as Editor-in-Chief of *Systems & Control Letters* and has been serving as Senior Editor for *Automatica and IEEE Transactions on Automatic Control*, as editor of two Springer book series—*Communications and Control Engineering and SpringerBriefs in Control, Automation and Robotics*—and has served as Vice President for Technical Activities of the IEEE Control Systems Society and as chair of the IEEE CSS Fellow Committee. Krstic has coauthored thirteen books on adaptive, nonlinear, and stochastic control, extremum seeking, control of PDE systems including turbulent flows, and control of delay systems.

Index

© Springer Nature Switzerland AG 2021
H. Sato, *Riemannian Optimization and Its Applications*,
SpringerBriefs in Control, Automation and Robotics,
https://doi.org/10.1007/978-3-030-62391-3

Printed in the United States
By Bookmasters